Carbon Management for a Sustainable Environment

Shelley W. W. Zhou

Carbon Management for a Sustainable Environment

 Springer

Shelley W. W. Zhou
Department of Civil & Environmental Engineering
Hong Kong University of Science and Technology
Clear Water Bay, Kowloon, Hong Kong

ISBN 978-3-030-35064-2   ISBN 978-3-030-35062-8   (eBook)
https://doi.org/10.1007/978-3-030-35062-8

This Springer imprint is published by the registered company Springer Nature Switzerland AG
The registered company address is: Gewerbestrasse 11, 6330 Cham, Switzerland

To my parents, whose love and support sustains me; and to my son Kwan and his generation, who hold the future of our planet in their hands.

Preface

The original purpose of writing this book was to summarize my 10 years of teaching at the Hong Kong University of Science and Technology and to present a readily available textbook for my students, especially those MSc students who haven't got the chance to take my course but doing the relevant carbon management research projects. But when I finished this book, I found it benefits not only students who are taking my courses but also readers from engineering and management consultants, corporate sustainability professionals, businessmen and government policy-makers who are all potential audience groups.

This book first introduces the basic concepts such as climate change, the relation between our daily activities and global warming, the impacts of climate change and why carbon footprint could be a parameter to measure the impacts. It then discusses how to measure carbon footprint and organizational greenhouse gas inventories. Most importantly, carbon management models have been introduced, which provide a holistic approach for companies or individuals to manage their carbon footprint. Moreover, the book exposes the readers to the best practice of carbon solutions, in terms of green buildings, smart transportation and waste management, which are the major areas from a business perspective, and the concept of carbon trading and offsetting, as a reduction means.

As the first batch of carbon consultant in this region, I have also presented various case studies to show how to develop and implement the overall carbon strategies and roadmap. I hope this book could provide sustainability practitioners and the students a comprehensive framework to conduct carbon management and could inspire the readers, no matter who they are, to study and practice further in the carbon management areas and equip them with climate change and carbon management concepts, carbon measurement skills and the competence and confidence to deliver them.

Clear Water Bay, Kowloon, Hong Kong   Shelley W. W. Zhou

Acknowledgements

I am grateful for all the friends and colleagues in encouraging me to start writing this book, persevere with it, and finally to publish it.

First and foremost, I would like to express my deep appreciation to Professor Irene M.C. LO at The Hong Kong University of Science and Technology (HKUST), who invited me to open a course on carbon management 10 years ago. Professor Lo has been my friend and mentor throughout my academic career, starting with my MPhil at HKUST in 1999.

My sincere thanks to Ir. Cary CHAN, Executive Director at Hong Kong Green Building Council, Ir. Colin CHUNG, Managing Director of Property and Buildings, and Sustainable Development and Environment, China, WSP, Ir. Albert LAI, CEO at Carbon Care Asia and James DONIVAN, Co-founder and CEO at ADEC Innovations, for delivering the guest lectures for my course and offering valuable business acumen and technical advices.

This book owes much to my writing advisor, "history professor" and travel companion, Robert BAXTER, who inspired me at the first place to start to work on it and shared and discussed with me throughout this project.

My special thanks to Zoe KENNEDY, Michael McCABE, Faith PILACIK and Amanda QUINN at Springer, and Batmanadan KARTHIKEYAN and Shobha KARUPPIAH at SPi Global, for their help and advice during the editorial and production stages in preparing this book.

Last, but not the least, I am extremely grateful to my parents for a lifetime of love, encouragement and support. Without them, this book would not have been possible.

Contents

List of Abbreviations/Symbols

AR5	Fifth Assessment Report
BPR	Business Process Reengineering
CDM	Clean Development Mechanisms
CERs	Certified Emission Reductions
CFCs	Chlorofluorocarbons
CFMP	Carbon Footprint Management Plan
CHP	Combined Heat and Power
CI	Carbon Intensity
COP	Conference of Parties
DDPP	Deep Decarbonization Pathway Project
DOEs	Designated Operational Entities
ETSs	Emission Trading Systems
GDP	Gross Domestic Product
GHG	Greenhouse Gas
GWP	Global Warming Potential
HFCs	Hydrofluorocarbons
IPCC	Intergovernmental Panel on Climate Change
ITP	International Tourism Partnership
JI	Joint Implementation
LCA	Life Cycle Assessment
mCM3	Modified Carbon Management Maturity Model
MOD	Mobility on Demand
MSW	Municipal Solid Waste
NDCs	Nationally Determined Contributions
OECD	Organisation for Economic Co-operation and Development
PCF	Product Carbon Footprint
PDD	Project Design Document
PFCs	Perfluorocarbons
RCPs	Representative Concentration Pathways
SDGs	Sustainable Development Goals
SF6	Sulphur Hexafluoride

TCFD	Taskforce on Climate-Related Financial Disclosure
TCM	Total Carbon Management
UNFCCC	United Nations Framework Convention on Climate Change
USDOT	United States Department of Transport
USEPA	United States Environmental Protection Agency
VCS	Verified Carbon Standard
VERs	Verified Emission Reductions

Chapter 1
Climate Change Basics

1.1 Let's Set the Scene

What do these two figures (Figs. 1.1 and 1.2) tell us?

- Growth domestic product (GDP) per capita has been grown substantially during the past two centuries.
- Economy grows at different rates for different economies. For instance, East Asia and Africa are far below the world average.
- World population has also increased dramatically, especially for the last century.

What questions we may have in our minds?

- What if the world continues economic growth dramatically?
- What if the poor countries, as they hope and rightly deserve, catch up with the high-income countries?
- What if population continues to grow?

With the economic growth, people's living quality and living conditions increase, which results in more production and consumption. Based on the growing trend from the above figures, i.e. in the business-as-usual scenario, it is anticipated there would be different levels of increased use of energy (i.e. majority from fossil fuels), depletion of natural resources, more emissions and pollutions and damage to the ecosystems, etc. The concept of planetary boundaries raised by a group of scientists at Stockholm Resilience Centre, asks the question in specificity:

- What are the major challenges coming from humanity's impact on the physical environment?
- Can we identify those challenges?
- Can we quantify them?
- Can we identify what would be safe limits for human activity so that we can begin rather urgently because we're late to this?

© Springer Nature Switzerland AG 2020
S. W. W. Zhou, *Carbon Management for a Sustainable Environment*,
https://doi.org/10.1007/978-3-030-35062-8_1

Average real GDP per capita across countries and regions
The measures are adjusted for inflation (at 2011 prices) and also for price differences between regions (multiple benchmarks allow for cross-country and regional income comparisons).

Fig. 1.1 World real GDP per capita (Roser 2018)

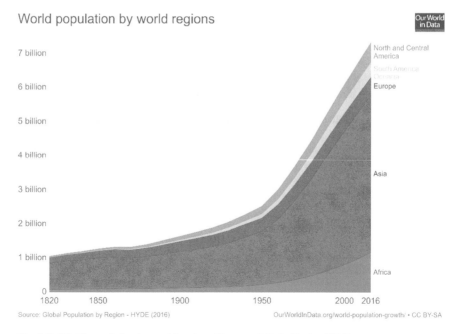

Fig. 1.2 World population by world regions (Roser and Ortiz-Ospina 2018)

Rockstrom et al. (2009) proposed the concept of the planetary boundary within which the Earth our planet and humanity can operate safely. The nine defined planetary boundaries as shown in Fig. 1.3 are climate change, ocean acidification, stratospheric ozone, biogeochemical nitrogen (N) cycle and phosphorus (P), global freshwater use, land system change, biological diversity lost, chemical pollution and atmospheric aerosol loading. Steffen et al. (2015) further approved that of these nine proposed boundaries, three (climate change, stratospheric ozone depletion and ocean acidification) might push the Earth system into a new state if crossed and they also have a pervasive influence on the remaining boundaries. Therefore, the ultimate aim of sustainable development is to redesign our technologies and our economic growth dynamics so that we can have economic improvement while staying within the planetary boundaries.

In this chapter, we will discuss the basics of climate change concepts, including its causes, the evidences and its consequences. We will then discuss the global effort

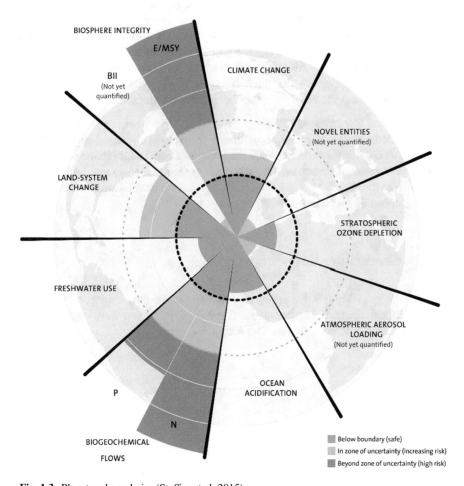

Fig. 1.3 Planetary boundaries (Steffen et al. 2015)

in tackling climate change and introduce the term of carbon footprint, the parameter by which we measure the impact of the climate change.

1.2 Greenhouse Effect and Greenhouse Gases

1.2.1 The Greenhouse Effect

The Earth's climate is driven by a continuous flow of energy from the sun. This energy arrives mainly in the form of visible light. About 30% is immediately scattered back into space, but most of the remaining 70% passes down through the atmosphere to warm the Earth's surface. The Earth must send this energy back out into space in the form of infrared radiation. While some of this infrared energy does radiate back into space, some portion is absorbed and re-emitted by water vapour and other greenhouse gases in the atmosphere. Greenhouse gases in the atmosphere block infrared radiation from escaping directly from the surface to space. Infrared radiation cannot pass straight through the air like visible light. Instead, most departing energy is carried away from the surface by air currents, eventually escaping to space from altitudes above the thickest layers of the greenhouse gas blanket. Figure 1.4 illustrates the basic processes behind the greenhouse effect.

The Enhanced Greenhouse Effect
The enhanced greenhouse effect, sometimes referred to as climate change or global warming, is the impact on the climate from the additional heat retained due to the increased amounts of carbon dioxide and other greenhouse gases that humans have released into the Earth's atmosphere since the Industrial Revolution. It is also called the anthropogenic greenhouse effect.

1.2.2 Greenhouse Gases

Greenhouse gas (GHG) is the gaseous constituent of the atmosphere, both natural and anthropogenic, that absorbs and emits radiation at specific wavelengths within the spectrum of infrared radiation. The main GHGs are:

- Water vapour
- Carbon dioxide
- Ozone
- Methane
- Nitrous oxide
- Halocarbons
- Other industrial gases

Apart from the industrial gases, all of these gases occur naturally. Together, they make up less than 1% of the atmosphere. This is enough to produce a natural

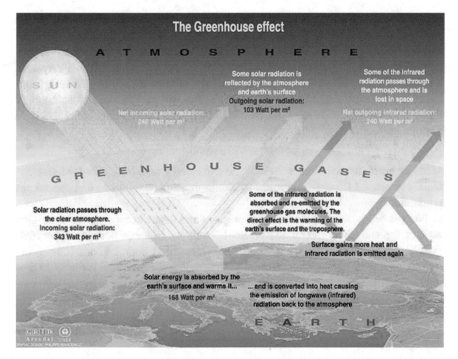

Fig. 1.4 An overview of the greenhouse effect. (From IPPC Working Group 1 contribution, Science of Climate Change, Second Assessment Report 1996)

greenhouse effect that keeps the planet some 30 °C warmer than it would otherwise be – essential for life as we know it.

Water Vapour

Water vapour is the largest contributor to the natural greenhouse effect. It is a naturally occurring GHG, which means human activity does not directly affect it. Nevertheless, water vapour has a so-called positive feedback to climate change – warmer air can hold more moisture – and models predict that a small global warming would lead to a rise in global water vapour levels, further adding to the enhanced greenhouse effect.

Carbon Dioxide

Carbon dioxide is currently responsible for over 60% of the enhanced greenhouse effect. Carbon dioxide produced by human activity enters the natural carbon cycle, such as burning of fossil fuel and deforestation. Even with half of humanity's carbon dioxide emissions being absorbed by the oceans and land vegetation, atmospheric levels continue to rise by over 10% every 20 years.

Aerosols

A second important human influence on climate is aerosols. These clouds of microscopic particles are not a greenhouse gas. In addition to various natural sources, they are produced from sulphur dioxide emitted mainly by power stations and by the smoke from deforestation and the burning of crop wastes. Aerosols settle out of the air after only a few days, but they are emitted in such massive quantities that they have a substantial impact on climate. Instead of global warming, aerosols may have a global dimming effect. Most aerosols cool the climate locally by scattering sunlight back into space and by affecting clouds. Aerosol particles can block sunlight directly and also provide seeds for clouds to form, and often these clouds also have a cooling effect.

Other GHGs
Methane currently contributes 20% of the enhanced greenhouse effect. Methane is mainly generated from agriculture, waste dumping and coal mining and natural gas production.

Nitrous oxide, a number of industrial gases and ozone contribute the remaining 20% of the enhanced greenhouse effect. Nitrous oxide levels have risen by 16%, mainly due to more intensive agriculture. While chlorofluorocarbons (CFCs) are stabilizing due to emission controls introduced under the Montreal Protocol to protect the stratospheric ozone layer, levels of long-lived gases such as hydrofluorocarbons (HFCs), perfluorocarbons (PFCs) and sulphur hexafluoride (SF6) are increasing.

1.3   Anthropogenic Evidence

1.3.1   Climate Change Evidence

The Intergovernmental Panel on Climate Change (IPCC) is the United Nations body for assessing the science related to climate change. The IPCC provides regular assessments of the scientific basis of climate change, its impacts and future risks and options for adaptation and mitigation. Since its inception in 1988, the IPCC has prepared five multivolume assessment reports. The Fifth Assessment Report (AR5) was released between September 2013 and November 2014. The IPCC is currently in its Sixth Assessment cycle. Figure 1.5 from IPCC AR5 (IPCC 2014a) illustrates the global warming is evident from observations of increases in global average air and ocean temperature, widespread melting of snow and ice and rising global mean sea level in the last 100 years.

Warming
- Each of the last three decades has been successively warmer at the Earth's surface than any preceding decade since 1850.
- The period from 1983 to 2012 was very likely the warmest 30-year period of the last 800 years in the Northern Hemisphere, where such assessment is possible and likely the warmest 30-year period of the last 1400 years.

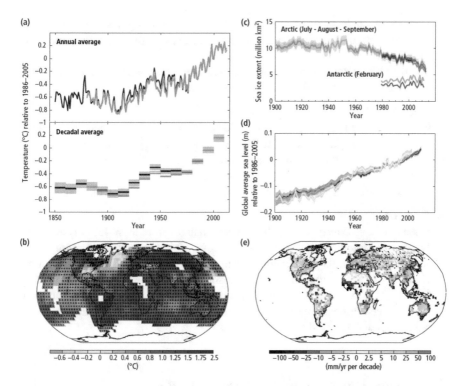

Fig. 1.5 Multiple observed indicators of a changing global climate system. (**a**) Observed globally averaged combined land and ocean surface temperature anomaly 1850–2012. (**b**) Observed change in surface temperature 1901–2012. (**c**) Sea ice extent. (**d**) Global mean sea level change 1900–2010. (**e**) Observed change in annual precipitation over land 1951–2010 (IPCC 2014a)

- The globally averaged combined land and ocean surface temperature data as calculated by a linear trend show a warming of 0.85 °C, over the period 1880–2012.
- The total increase between the average of the 1850–1900 period and the 2003–2012 period is 0.78 °C, based on the single longest dataset available.

Sea Ice Extent
- Over the last two decades, the Greenland and Antarctic ice sheets have been losing mass.
- Glaciers have continued to shrink almost worldwide.
- The extent of Northern Hemisphere snow cover has decreased since the mid-twentieth century by 1.6% per decade for March and April, and 11.7% per decade for June, over the 1967–2012 period.

Sea Level Rising
- Over the period 1901–2010, global mean sea level rose by 0.19 m.

- The rate of sea level rise since the mid-nineteenth century has been larger than the mean rate during the previous two millennia, and the rise will continue to accelerate.
- It is very likely that the mean rate of global averaged sea level rise was 1.7 mm/yr between 1901 and 2010 and 3.2 mm/yr between 1993 and 2010.

Extreme Weather
- It is very likely that the number of cold days and nights has decreased and the number of warm days and nights has increased on the global scale.
- More droughts, heavy precipitation, heat waves and the intensity of tropical cyclones.
- More powerful storms and hotter, longer dry periods have been observed.

Shift in Natural World
- Scientists have observed climate-induced changes in at least 420 physical processes and biological species or communities. Changes include migratory birds arriving earlier in the spring and leaving later in the autumn, a lengthening by 10.8 days of the European growing season for controlled mix-species gardens from 1959 to 1993, earlier springtime reproduction for many birds and amphibians and the northward movement of cold-sensitive butterflies, beetles and dragonflies (UNEP and UNFCCC 2002).

1.3.2 How Human Activities Produce GHGs

Based on IPCC's Fourth Assessment Report (AR4) (IPCC 2007), it was concluded that most of the observed increase in global mean temperature since the mid-twentieth century is "very likely" due to the observed increase in anthropogenic GHG concentrations; it is then emphasized on the AR5 (IPCC 2014a) that "extremely likely" (certainty >95%) that human influence is the dominant cause of the observed warming since the mid-twentieth century.

Figure 1.6 shows indicators of the human influence on the atmosphere during the Industrial Era. It can be seen that all three records show effects of the large and increasing growth in anthropogenic emissions during the Industrial Era.

Use of Fossil Fuels
- The supply and use of fossil fuels account for about 80% of anthropogenic carbon dioxide (CO_2) emissions, 20% of the methane (CH_4) and a significant quantity of nitrous oxide (N_2O).
- Coal, oil and natural gas are the main energy supply for electricity generation, transportation, heating and power supply.
- Extracting, processing, transporting and distributing fossil fuels also release greenhouse gases.

Deforestation
- Deforestation is the second largest source of carbon dioxide.

- Due to the photosynthesis, trees and forest are the natural carbon sink, which can absorb carbon dioxide. Clearing forest for other land use, burning or decomposing trees, would reduce the carbon sink, which equals to increase the carbon emission on the mass balance point of view.

Cement Production

- Globally, approximately 3.4% of global CO_2 emissions are from cement industrial sources.
- CO_2 is released when coal and petroleum coke are used to fuel the kilns for clinker production, as well as during the chemical process of calcination of limestones.

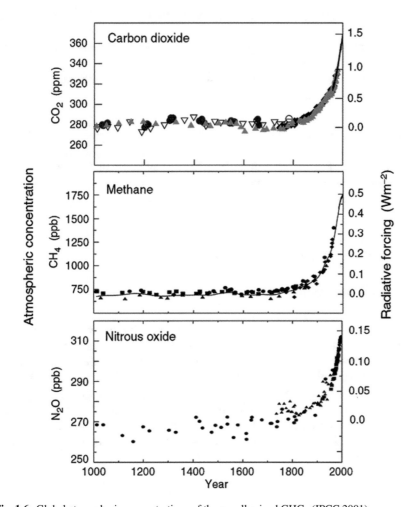

Fig. 1.6 Global atmospheric concentrations of three well-mixed GHGs (IPCC 2001)

Livestock
- Livestock account for 30% of the methane emissions from human activities.
- Methane is produced in the animals' digestive tracts as well as decomposition of animal manures.

Agriculture
- Wetland or paddy rice farming produces roughly one-fifth to one-quarter of global methane emissions from human activities.
- Fertilizer use increases nitrous oxide emissions. The nitrogen contained in many mineral and organic fertilizers and manures enhances the natural processes of nitrification and denitrification that are carried out by bacteria and other microbes in the soil. These processes convert some nitrogen into nitrous oxide.

Waste Management
- Landfill gas (mainly in methane) will be generated at landfill site of municipal solid waste (MSW).
- Biogas (methane) will be generated during the anaerobic digestion of the organic waste (e.g. food waste).
- Unless methane from both cases above is captured and utilized, it will then release into the atmosphere.

Industries
- Industry has created a number of long-lived and potent greenhouse gases for specialized uses.
- CFCs, HFCs, PFCs – used in many industries as refrigerants, blowing agents, propellants in medicinal applications and degreasing solvents.
- SF6 – used as an electric insulator, heat conductor and freezing agent.

1.4   Climate Change Consequences

1.4.1   Climate Risks and Impacts

Continued emission of greenhouse gases will cause further warming and long-lasting changes in all components of the climate system, increasing the likelihood of severe, pervasive and irreversible impacts for people and ecosystems. It will threaten the basic elements of life for people around the world – access to water, food production, health and use of land and the environment. Figure 1.7 illustrates the climate impacts with more anthropogenic GHGs (Stern 2006). The climate risks and impacts are assessed based on the projected future climate changes of different mitigation scenarios, i.e. Representative Concentration Pathways (RCPs) (IPCC 2014a).

Warming
- The global mean surface temperature change for the period 2016–2035 relative to 1986–2005 is similar for the four RCPs and will likely be in the range 0.3–0.7 °C.
- Projected increase in global mean surface temperature for the mid- and late twenty-first century, relative to the 1986–2005 period, is in the range of 1.0–2.0 °C and 1.0–3.7 °C, respectively.
- The Arctic region will continue to warm more rapidly than the global mean.

Sea Level Rising
- Global mean sea level will continue to rise during the twenty-first century.
- Projected change in global mean sea level rise for the mid- and late twenty-first century, relative to the 1986–2005 period, is in the range of 0.24–0.3 m and 0.40–0.63 m, respectively.
- Sea level rise will not be uniform across regions. By the end of the twenty-first century, it is very likely that sea level will rise to more than about 95% of the ocean area.
- About 70% of the coastlines worldwide are projected to experience sea level change within ±20% of the global mean.

Water Resources
- In many regions, changing precipitation and melting snow and ice are altering hydrological systems, affecting water resources in terms of quantity and quality.
- In presently dry regions, the frequency of droughts will likely increase by the end of the twenty-first century. In contrast, water resources are projected to increase at high latitudes.
- The interaction of increased temperature; increased sediment, nutrient and pollutant loadings from heavy rainfall; increased concentrations of pollutants during droughts; and disruption of treatment facilities during floods will reduce raw water quality and pose risks to drinking water quality.
- Climate change is projected to reduce renewable surface water and groundwater resources in most dry subtropical regions, intensifying competition for water among sectors.

Agriculture and Food Security
- All aspects of food security are potentially affected by climate change, including food production, access, use and price stability.
- Due to projected climate change by the mid-twenty-first century and beyond, global marine species redistribution and marine biodiversity reduction in sensitive regions will challenge the sustained provision of fisheries productivity and other ecosystem services.
- For wheat, rice and maize in tropical and temperate regions, climate change without adaptation is projected to negatively impact production at local temperature increases of 2 °C or more above late-twentieth-century levels.

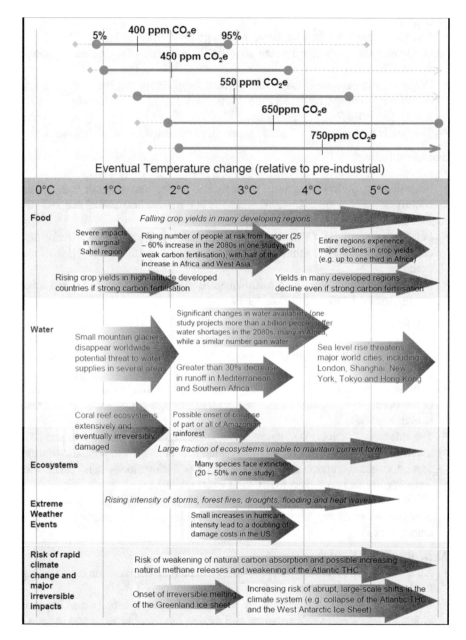

Fig. 1.7 Climate impacts (Stern 2006)

- Global temperature increases of ~4 °C or more above late-twentieth-century lev-els, combined with increasing food demand, would pose large risks to food secu-rity globally.

Biodiversity
- The resilience of many ecosystems is likely to be exceeded this century by an unprecedented combination of climate change, associated disturbances (e.g. flooding, drought, wildfire, insects, ocean acidification) and other global change drivers (e.g. land-use change, pollution, fragmentation of natural systems, over-exploitation of resources).
- Approximately 20–30% of plant and animal species assessed so far are likely to be at increased risk of extinction if increases in global average temperature exceed 1.5–2.5 °C.
- Many terrestrial, freshwater and marine species have shifted their geographic ranges, seasonal activities, migration patterns, abundances and species interac-tions in response to ongoing climate change.

Human Health
- Increasing burden from malnutrition and diarrhoeal, cardiorespiratory and infec-tious diseases
- Increased morbidity and mortality from heat waves, floods and droughts
- Changed distribution of some disease vectors
- Substantial burden on health services

Examples of Superyphoon Impacts in Hong Kong

The 1972 rainfall-triggered Po Shan Road landslide, Hong Kong, which killed 67 people when a 12-story apartment building was destroyed by the 50,000 m 3 flowslide. (Photo by Geotechnical Control Office, Hong Kong Government) (Schuster and Highland 2007)

After Mangkhut, the most intense storm in Hong Kong's history, hit Hong Kong, *South China Morning Post*, a local newspaper said on 18 September 2018 that "death tolls are down but damage bills are up as pollution and development pressures leave the region more exposed to disaster". Chuck Watson, a disaster modeller for Enki Research in the US city of Savannah, initially estimated that mainland China's losses from Mangkhut could be as much as US$100 billion. That was on top of the US$20 billion in damage inflicted on Hong Kong (http://www.enkiops.org/blog/page/2/).

1.4.2   Adaptation and Mitigation

Adaptation
The process of adjustment to actual or expected climate and its effects. In human systems, adaptation seeks to moderate or avoid harm or exploit benefi-cial opportunities. In some natural systems, human intervention may facilitate adjustment to expected climate and its effects (IPCC 2014a).

Vulnerability
The propensity or predisposition to be adversely affected. Vulnerability encompasses a variety of concepts and elements including sensitivity or sus-ceptibility to harm and lack of capacity to cope and adapt (IPCC 2014a).

Adaptation and mitigation are two complementary strategies for responding to climate change. Adaptation is the process of adjustment to actual or expected climate and its effects in order to either lessen or avoid harm or exploit beneficial opportunities. Mitigation is the process of reducing emissions or enhancing sinks of greenhouse gases, so as to limit future climate change. Both adaptation and mitigation can reduce and manage the risks of climate change impacts.

Example: Netherlands' Adaptation in Defending Flooding

The Netherlands represents one of the most vulnerable regions to sea level rise in the world. It is a land of waterways, and 26% is below sea level, with 60% of its people and 70% of GDP earned in flood-risk areas. There is deep experience of what it takes to deal with flooding, in both financial and human terms. The Netherlands has been fighting back water for more than 1000 years, when farmers built the first dykes and an important paradigm shift is from flood protection to flood management, where Dutch people do not fight with water, but continue living with water with the adaptive infrastructures. One such example is 22 acres of reclaimed canals just outside of Rotterdam, an area called the Eendragtspolder, which serves to collect water from the Rotte River Basin when the nearby Rhine River overflows, which it is anticipated to do every 10 years due to climate change. The Eendragtspolder is also home to bike paths, water sports and a brand-new rowing course, where the World Rowing Championships were staged in 2016 summer (Bart de Jong 2016).

Example: Hong Kong's Major Mitigation Measures

Electricity Supply	Buildings	Transport	Waste
Reduce coal usage in fuel mix through using more natural gas, nuclear electricity and RE	Drive energy saving through various means: • New buildings – better design and construction; • Existing buildings – re-commissioning; auditing and better management; and retrofitting; • Change property management's and inhabitants' behaviour based on energy demand monitoring and forecast ; • Influence inhabitants to buy energy efficient electrical products	• Make public transport primary choice for mobility • Expand rail options and services • Improve rail operation energy efficiency • Improve vehicle fuel efficiency • Testing low-carbon and zero emissions franchised bus technologies	• Recover and use landfill gas • Recover energy from sludge treatment • Develop waste-to-energy treatment for organic and yard waste and municipal solid waste

Environment Bureau (2015)

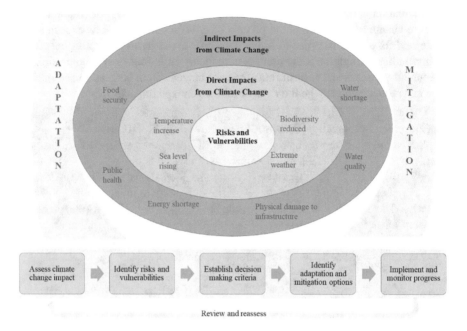

Fig. 1.8 Climate change risk framework and management process

Figure 1.8 shows the overall framework that could be used for business decision-makers. The main steps of the climate risk management process include:

- Assessing climate change impacts, both direct and indirect ones
- Identifying the climate risks and its vulnerabilities
- Establishing the decision-making criteria
- Identifying the possible adaptation and mitigation options
- Implementing the decisions and monitoring the results
- Reviewing the progress and reassessing the risks and improving continuously

Although this book is mainly focused on discussing the mitigation measures, i.e. how companies could reduce their carbon emission through carbon management, climate resilience is of the same importance, if not more, to companies in different sectors. Adaptation measures and how to build climate resilience will be discussed briefly in Chap. 5, as part of the total carbon management.

1.5   The Climate Change Convention

1.5.1   UNFCCC

The United Nations Framework Convention on Climate Change (UNFCCC) is the foundation of global efforts to combat global warming. Opened for signature in 1992 at the Rio Earth Summit, the Convention sets out some guiding principles. The precautionary principle says that the lack of full scientific certainty should not be used as an excuse to postpone action when there is a threat of serious or irreversible damage. The principle of the "common but differentiated responsibilities" of states assigns the lead in combating climate change to developed countries.

- Both developed and developing countries accept a number of general commitments.
- Industrialized countries undertake several specific commitments.
- The richest countries shall provide new and additional financial resources and facilitate technology transfer.

The supreme body of the Convention is the Conference of the Parties (COP). It held its first meeting (COP 1) in Berlin in 1995 and continues to meet on a yearly basis unless the Parties decide otherwise. The COP's role is to promote and review the implementation of the Convention. The COP can adopt new commitments through amendments and protocols to the Convention. The two most important ones are Kyoto Protocol and Paris Agreement. Figure 1.9 highlights the main events on UNFCCC timeline.

1.5.2   Kyoto Protocol

Adopted by consensus at the third session of the Conference of the Parties (COP 3) in December 1997, Kyoto Protocol contains legally binding emissions targets for Annex I (industrialized) countries. Most members of the Organisation for Economic Co-operation and Development (OECD) plus the states of Central and Eastern Europe are known collectively as Annex I countries.

The developed countries are to reduce their collective emissions of six key greenhouse gases (i.e. CO_2, CH_4, N_2O, HFCs, PFCs and SF_6) at least 5%.

- Each country's emissions target must be achieved by the period 2008–2012.
- Cuts in the three most important gases, CO_2, CH_4 and N_2O, will be measured against a base year of 1990.
- Cuts in three long-lived industrial gases, HFCs, PFCs and SF_6, can be measured against either a 1990 or 1995 baseline.
- Countries will have some flexibility in how they make and measure their emission reductions.

1.5.3   From Kyoto to Paris

Under the Kyoto Protocol, 38 countries called Annex B-1997 countries that account for 39% of 2010 global GHG emissions committed to a 5% decrease in their emissions between 2008 and 2012 in comparison with their base-year emissions, i.e.1990 or 1995, for some GHGs as discussed above. With the non-ratification of the Kyoto Protocol by the USA and the withdrawal of Canada in 2011, the commitments of the 36 remaining countries so-called Annex B-2012 countries accounted for 24% of

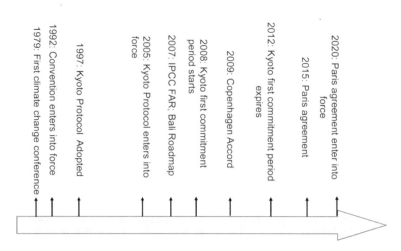

Fig. 1.9 UNFCCC timelines

global GHG emissions in 2010. The study published in 2016 showed that the average annual aggregated GHG emissions of Annex B-2012 countries in the first commitment period, i.e. 2008–2012, were 24% below the base-year emissions, while their aggregate target was only 4% reduction. The overall Kyoto target has thus been overachieved by 2.4 GtCO$_2$e (Shishlov et al. 2016).

However, excluding the participants of two main emitters, the USA and China, and other developing countries, Kyoto Protocol could not bring the global emission down in such a short period of time. It has taken a long time in UN climate conferences to discuss a post-2012 target and a legally binding agreement that could involve all the nations. COP 15 in 2009 at Copenhagen was a culmination of a 2-year negotiating process under the Bali Roadmap for a post-2012 framework, and at COP-18 in 2012 at Doha, Qatar, an amendment to the Kyoto Protocol was adopted which ultimately created a second commitment period among member Parties, and it extended the Protocol from 1 January 2013 until 31 December 2020. Another three years passed till the COP 21 sustainable development summit, held in Paris, all UNFCCC participants signed the "Paris Agreement" effectively replacing the Kyoto Protocol.

1.5.4 Paris Agreement

In December 2015 at COP 21 in Paris, 195 countries adopted the first-ever universal, legally binding global climate deal Paris Agreement. The agreement sets out a global action plan to put the world on track to avoid dangerous climate change by limiting global warming to well below 2 °C above pre-industrial levels. The agreement is due to enter into force in 2020. It provides a bridge between today's policies and climate neutrality before the end of the century.

Mitigation

Governments agreed:

- The long-term goal of limiting global temperature increase well below 2 °C
- Urging efforts to limit the increase to 1.5 °C, since this would significantly reduce risks and the impacts of climate change
- On the need for global emissions to peak as soon as possible, recognizing that this will take longer for developing countries
- To undertake rapid reductions thereafter in accordance with the best available science
- Establishing binding commitments by all Parties to make "nationally determined contributions" (NDCs) and to pursue domestic measures aimed at achieving them

It should be noted that while Kyoto Protocol is mainly a top-down approach, universally participated Paris Agreement has an innovative hybrid structure blending both top-down and bottom-up elements.

Transparency and Global Stocktake

Governments agreed:

- To submit new NDCs every 5 years, with the clear expectation that they will "represent a progression" beyond previous ones
- To report regularly on their emissions and "progress made in implementing and achieving" their NDCs and to undergo international review
- To track progress towards the long-term goal through a robust transparency and accountability system

Adaptation

Governments agreed:

- To strengthen societies' ability to deal with the impacts of climate change
- To provide continued and enhanced international support for adaptation to developing countries

Support

- Reaffirm the binding obligations of developed countries under the UNFCCC to support the efforts of developing countries while for the first time encouraging voluntary contributions by developing countries too.
- Extend the current goal of mobilizing $100 billion a year in support by 2020 through 2025, with a new, higher goal to be set for the period after 2025.

China's Nationally Determined Actions by 2030

- To achieve the peaking of carbon dioxide emissions around 2030 and making best efforts to peak earlier
- To lower carbon dioxide emissions per unit of GDP by 60–65% from the 2005 level
- To increase the share of non-fossil fuels in primary energy consumption to around 20%
- To increase the forest stock volume by around 4.5 billion cubic metres on the 2005 level

1.6   Sustainable Development Goals (SDGs)

In the same year of Paris climate conference, actually just 3 months before it, the UN General Assembly adopted the outcome document named *Transforming our world: the 2030 Agenda for Sustainable Development*, in short form known as 2030 Agenda. It set out 17 Sustainable Development Goals (SDGs) and 169 targets to be reached by 2030 with an aim to eliminate poverty and bring the world on a path to sustainable development.

As shown in Fig. 1.10, most the first nine goals are about the economic dimension – ending poverty and ensuring access to basic needs like food, health and education, water and cleaner energy services. SDGs 5, 10 and 16 are especially about social inclusion – SDG 5 on gender equality, SDG 10 on reducing inequality and SDG 16 on peace and justice of the society. Then goals 11–15 are primarily the environmental objectives – sustainable cities, sustainable production and consumption, biodiversity and sustainable ecosystem both in the oceans and on land and SDG 13 on climate change. Finally, SDG 17 says all these economic, social and environmental goals should be achieved in collective effort by partnership.

Now SDG 13. Based on 2030 Agenda, it says "take urgent action to combat climate change and its impacts" with three targets:

13.1 Strengthen resilience and adaptive capacity to climate-related hazards and natural disasters in all countries.
13.2 Integrate climate change measures into national policies, strategies and planning.
13.3 Improve education, awareness-raising and human and institutional capacity on climate change mitigation, adaptation, impact reduction and early warning.

There is an asterisk next to the goal 13 on the UN's document 2030 Agenda; with a footnote it acknowledges UNFCCC as the primary international, intergovernmental forum for negotiating the global response to climate change. It thus does not set

Fig. 1.10 Sustainable Development Goals

specific, measurable targets for mitigation or adaptation, leaving that task to the Paris Agreement.

So how climate actions formulated in NDCs as described by Paris Agreement correspond to SDGs and particularly SDG 13? NDC-SDG Connections is a joint initiative of the German Development Institute (DIE) and Stockholm Environment Institute (SEI). They connect climate action to each of 17 SDGs. Here are their findings:

- Although only SDG 13 is explicitly about tackling climate change, all aspects of sustainable development will be impacted by the effects of a changing climate and by the actions of countries under the Paris Agreement on Climate Change, embodied in countries' NDCs.
- While all NDCs are inherently connected to SDG 13, only 6% of activities directly correspond to this goal. A reason for this is that the SDG targets are relatively narrow and focus on adaptive capacity, policy mainstreaming and education and awareness. Other relevant issues are related to climate mitigation or covered by, for example, SDG 7 on cleaner energy.
- At the level of targets, the NDC activities most frequently relate to target 13.2 and target 13.3. Target 13.1 is not equally prominent.
- In terms of climate actions, countries address mainly issues like awareness-raising on climate impacts. Many SDG 13-relevenat NDC activities refer to themes highly relevant for other SDGs, including energy (SDG 7) and education (SDG 4) as well as resilience and disaster risk management, which appear under several SDGs, most notably SDG 2 (zero hunger) and SDG 11 (sustainable cities).

More information can be found from https://klimalog.die-gdi.de/ndc-sdg/sdg/13.

In this book, I will discuss what Paris Agreement means to business sectors and individual organizations:

– How companies could set the carbon reduction target to contribute its share to Paris Agreement
– How companies could take deep decarbonization pathway
– How they could build up their climate resilience and capacity

Relevant SDGs will also be addressed in Chap. 7 on reduction solutions, especially for building sector.

1.7 Carbon Footprint Concept

In this chapter so far, we have discussed the enhanced greenhouse effects from anthropogenic GHGs generated by human's activities, the evidence of the climate change issues and the impacts and consequences projected by IPCC, which is summarized in Fig. 1.11. The question to us next is "Can we measure the impacts? If yes, how?"

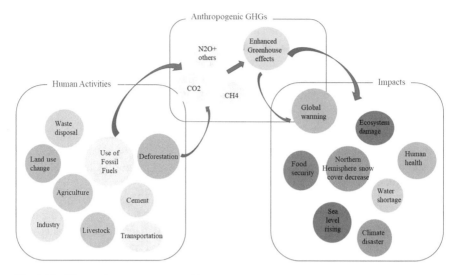

Fig. 1.11 Climate change processes and impacts

Carbon Footprint is a parameter that measures the impact human activities have on the environment in terms of the amount of greenhouse gases produced, measured in unit of carbon dioxide equivalent (CO_2e).

In this book, same as defined in GHG Protocol (WBCSD and WRI 2010), carbon footprint measures six greenhouse gases covered by the Kyoto Protocol – carbon dioxide (CO_2), methane (CH_4), nitrous oxide (N_2O), hydrofluorocarbons (HFCs), perfluorocarbons (PFCs) and sulphur hexafluoride (SF_6) (Table 1.1).

Global warming potential (GWP) is a measure of how much a given mass of GHG is estimated to contribute to global warming. It is a relative scale which compares the gas in question to that of the same mass of CO_2 (whose GWP is by convention equal to 1). GWP is a factor of:

- Radiative efficiency (heat-absorbing ability) of each gas relative to that of CO_2
- Decay rate of each gas (the amount removed from the atmosphere over a given number of years) relative to that of CO_2 (Table 1.2)

For GHGs other than CO_2:
 CO_2e = mass emitted × GWP (100) of substance
For fossil fuel/power consumption:
 CO_2e = unit of consumption × emission factor (EF)

Table 1.3 shows some activities in Hong Kong that emit 1 tonne of carbon footprint. It tells us that carbon footprint is the parameter used for measurement and for comparison, from which we could get a lower carbon solution. For instance, when we talk about food, to consume the same amount of food, beef has a much higher carbon emission than vegetables.

Table 1.1 Six Kyoto gases

Symbol	Name	Common sources
CO_2	Carbon dioxide	Fossil fuel combustion, forest clearing, cement production, etc.
CH_4	Methane	Landfills, production and distribution of natural gas and petroleum, fermentation from the digestive system of livestock, rice cultivation, fossil fuel combustion, etc.
N_2O	Nitrous oxide	Fossil fuel combustion, fertilizers, nylon production, manure, etc.
HFCs	Hydrofluorocarbons	Refrigeration gases, aluminium smelting, semiconductor manufacturing, etc.
PFCs	Perfluorocarbons	Aluminium production, semiconductor industry, etc.
SF_6	Sulphur hexafluoride	Electrical transmissions and distribution systems, circuit breakers, magnesium production, etc.

Table 1.2 Global warming potential

	GWP 100-year time horizon		
GHGs	SAR (IPCC 1995)[a]	TAR (IPCC 2001)	AR4 (IPCC 2007)
CO_2	1	1	1
CH_4	21	23	25
N_2O	310	296	298
HFC-134a	1,300	1,300	1,430
PFC-14	6,500		7,390
SF_6	23,900	22,200	22,800

[a]For no specific reason, the following examples of calculation in this book will use the GWP based on SAR (IPCC 1995)

Table 1.3 One tonne of carbon footprint in Hong Kong's perspective

Activity	Activity data[a]	
Electricity consumption	Hong Kong Island	1,250 kWh
	Kowloon	2,000 kWh
Transportation	Driving	600 litre gasoline
	Taxi	8,264 km (~ HK$ 47 K)
	MTR	128,205 km (~ HK$ 87 K)
Food consumption	Beef	37.5 kg
	Chicken	357 kg
	Vegetables	20 tonnes
Paper waste	Landfilling	208 kg
Tree[b]		44 trees

[a]The data in this table is the estimated data for reference only
[b]Here means as a carbon sink, 44 trees can absorb 1 tonne of carbon emission every year

References

Bart de Jong (2016) Flood defense in the Netherlands: a story of adaptation, presented at 2016DC flood summit at Gallaudet University, September 8, 2016

Environment Bureau (2015) Hong Kong climate change report 2015. Environment Bureau in collaboration with Development Bureau, Transport & Housing Bureau, Commerce & Economic Development Bureau, Food & Health Bureau, Security Bureau. November 2015

IPCC (1995) Climate Change 1995: The Science of Climate Change. Contribution of WGI to the Second Assessment Reportof the Intergovernmental Panel on Climate Change. Edited by J.T. Houghton, L.G. Meira Filho, B.A. Callander, N. Harris, A. Kattenberg and K. Maskell

IPCC (2001) Climate Change 2001: Synthesis Report. A Contribution of Working Groups I, II, and III to the Third Assessment Report of the Integovernmental Panel on Climate Change [Watson, R.T. and the Core Writing Team (eds.)]. Cambridge University Press, Cambridge, United Kingdom, and New York, NY, USA, 398 pp

IPCC (2007) Climate Change 2007: Synthesis Report. Contribution of Working Groups I, II and III to the Fourth Assessment Report of the Intergovernmental Panel on Climate Change [Core Writing Team, Pachauri, R.K and Reisinger, A. (eds.)]. IPCC, Geneva, Switzerland, 104 pp.

IPCC (2014a) Climate change 2014: synthesis report. Contribution of working groups I, II and III to the fifth assessment report of the intergovernmental panel on climate change, IPCC, Geneva, Switzerland, 151 pp

Roser M (2018) Economic growth. Published online at OurWorldInData.org. Retrieved from: https://ourworldindata.org/economic-growth [online resource]

Roser M, Ortiz-Ospina E (2018) World population growth. Published online at OurWorldInData. org. Retrieved from: https://ourworldindata.org/world-population-growth [online resource]

Rockstrom J, Steffen W, Noone K et al (2009) Planetary boundaries: exploring the safe operating space for humanity. Ecol Soc 14(2):32. [online] URL: http://www.ecologyandsociety.org/vol14/iss2/art32/

Schuster RL, Highland LM (2007) The third Hans Cloos lecture. Urban landslides: socioeconomic impacts and overview of mitigative strategies. Bull Eng Geol Environ 66:1–27

Shishlov I, Morel R, Bellassen V (2016) Compliance of the parties to the Kyoto protocol in the first commitment period. Clim Pol 16(6):768–782

Steffen W, Richardson K, Rockstrom J, Cornell SE et al (2015) Planetary boundaries: guiding human development on a changing planet. Science 347(6223):1259855. https://doi.org/10.1126/science.1259855

Stern N (2006) Stern review: the economics of climate change. https://webarchive.nationalarchives.gov.uk/+/http://www.hm-treasury.gov.uk/sternreview_index.htm

UNEP and UNFCCC (2002) Climate change information kit. Updated in July 2001 based on the IPCCs Climate Change: 2001 assessment report and current activities under the UN Framework Convention on Climate Change. Published by the United Nations Environment Programme and the Climate Change Secretariate (UNFCCC) and sponsored by UNEP, the UN Development Programme, the UN Department of Economic and Social Affairs, the UN Institute for Training and Research, the World Meteorological Organization, the World Health Organization, and the UNFCCC. Edited by Michael Williams

World Business Council for Sustainable Development and World Resources Institute (2010) The greenhouse gas protocol, a corporate accounting and reporting standard, revised version. https://ghgprotocol.org/corporate-standard

Chapter 2
Carbon Footprint Measurement

2.1 Why and What

From the above discussion, we know that carbon footprint is a parameter that measures the GHG emissions from the human's activities. In this chapter, we will further discuss how to measure this parameter or, in other words, how to prepare a GHG inventory. What we measure is "something's" carbon footprint and that "something" could be a building, a hotel, a company, a factory, a production line, a product, an event, an exhibition, etc., and we only focus on six Kyoto gases, as we mentioned in Sect. 1.7.

Why We Measure?

In 2006, Carbon Trust, a not-for-profit UK organization, helped PepsiCo to measure the carbon footprint of its Walkers crisps. Carbon Trust has gone through Walkers' supply chain, measured the direct energy consumption at each life cycle stage of a packet of crisps and calculated the carbon footprint for each stage, the contribution of which are listed below:

1. Raw materials: potatoes, sunflowers and seasoning and packaging (40%)
2. Manufacture: producing crisps from potatoes (30%)
3. Packaging (18%)
4. Distribution (10%)
5. Disposal of the empty packs (2%)

Raw materials and manufacturing are the largest sources (around 70%) of emissions across the crisps' life cycle. It was also found during the study that potatoes were purchased by weight, and Walkers paid a price per tonne of potatoes to farmers. Thus, in order to get more money, farmers stored their potatoes in artificially humidified warehousing shed, where humidified atmosphere increased water content of potatoes. Humidifiers used large amounts of energy and generated significant emissions on the farm, while extra moisture in potatoes also

© Springer Nature Switzerland AG 2020
S. W. W. Zhou, *Carbon Management for a Sustainable Environment*,
https://doi.org/10.1007/978-3-030-35062-8_2

needed more energy (around 10%) in frying at the manufacturing site. Walkers then engaged with its potato suppliers to reduce emissions through better agricultural and storage practices. It also scoured its own production facilities for opportunities to cut energy use, resulting in 33% reduction in energy use per kg crisps (Carbon Trust 2006, 2008).

This example tells us that carbon footprint measurement not only gives us the carbon footprint data (e.g. 75 g CO_2e per packet of crisps); it also helps to identify the hotspots where and how to reduce the emission (e.g. between 2007 and 2009, Walkers reduced its carbon footprint by 7%, creating an overall saving of 4,800 tonnes of CO_2e), which is our ultimate purpose in managing our carbon footprint to reduce the induced climate impacts. Measurement is the first step of management and also the most important step before reduction. We need to understand where we are, before we know which direction to take and how to move forwards. Global climate change is one of the most challenging issues facing policy makers and industry today. We need scientific measurements to improve our understanding of this global issue, which provides the foundation to deliver policies for mitigating climate change, and to accelerate the development of low carbon technologies.

2.2 International and Local Standards

Table 2.1 lists different international standards for carbon footprint measurement and reporting for different applications. For organizations, the commonly used standards are ISO 14064-1 (ISO 2018a) and GHG Protocol (WBCSD/WRI 2010). In Australia, the UK, the USA and Hong Kong, both standards are adopted, and additional guidance on local situations, such as local emission factors, is further added into the local or national standards, which is summarized in Table 2.2.

Table 2.1 International standards for carbon footprint measurement

Application	International standards
Organization	ISO 14064-1: Greenhouse Gases – Part 1: Specification with guidance at the organization level for quantification and reporting of greenhouse gas emissions and removals
	WBCSD/WRI: The GHG Protocol Corporate Accounting and Reporting Standard
Projects	ISO 14064-2: Greenhouse Gases – Part 2: Specification with guidance at the project level for quantification, monitoring and reporting of greenhouse gas emission reductions or removal enhancements
	WBCSD/WRI: The GHG Protocol for Project Accounting
Products and services	ISO 14067: Greenhouse gases – Carbon footprint of products – Requirements and guidelines for quantification
	Publicly Available Specification – PAS 2050 Specification for the assessment of the life cycle greenhouse gas emissions of goods and services
	WRI/WBCSD: The Product Life Cycle Accounting and Reporting Standard

Table 2.2 Local standards or tools for carbon footprint measurement

Application	National standards
Organizations	Australia Department of the Environment and Energy: The National Greenhouse Accounts (NGA) Factors
	UK DEFRE/DECC Guidance on how to measure and report your greenhouse gas emissions
	USEPA Climate Leaders Greenhouse Gas Inventory Protocol
Buildings	HKEPD and EMSD Guidelines: Guidelines to Account for and Report on Greenhouse Gas Emissions and Removals for Buildings (Commercial, Residential or Institutional Purposes) in Hong Kong

2.3   Corporate Carbon Audit: Defining the Boundaries

The first step in the carbon footprint measurement or development of a GHG emission inventory is to define the boundaries of the measurement or the inventory. These boundaries refer to the coverage and extent that will be taken into account for the measurement, i.e. they determine what should be included and what can be excluded. Organizational boundaries define the operations, facilities and entities that are to be included in the inventory, while operational boundaries identify emission sources and categorize the emissions resulting either directly or indirectly from the organization's operations, facilities and entities. In this section, we will discuss how to define these two boundaries, mainly based on the GHG corporate protocol.

2.3.1   Organizational Boundary Setting

Business operations vary in their legal and organizational structures and include wholly owned operations, subsidiaries, incorporated and non-incorporated joint ventures, etc. Organizational boundaries determine which operations, facilities and entities owned or controlled by the reporting company should be reported, depending on the approach chosen. The company should consistently apply the selected approach to define those businesses and operations that constitute the company for the purpose of accounting and reporting GHG emissions. For corporate reporting, two approaches can be used to define the organizational boundaries: the equity share approach and the control approach (Fig. 2.1).

Equity Share Approach: A company accounts for GHG emissions from operations according to its share of equity in the operation.

- The equity share reflects the percentage of economic interest.
- Equity share normally is the same as the ownership percentage.
- If you choose the equity share approach, you must report all emissions sources that are wholly owned and partially owned according to your entity's equity share in each.

Control Approach: A company accounts for 100 percent of the GHG emissions from operations over which it has control. It does not account for GHG emissions

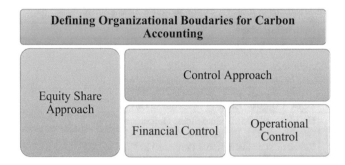

Fig. 2.1 Summary of different approaches to define the organizational boundaries

from operations in which it owns an interest but has no control. Control can be defined in either financial or operational terms. When using the control approach to consolidate GHG emissions, companies shall choose between either the operational control or financial control criteria.

- An entity has **operational control** over an operation if the entity or one of its subsidiaries has the full authority to introduce and implement its operating policies. The entity that holds the operating licence for an operation typically has operational control.
- An entity has **financial control** over an operation if the entity has the ability to direct the financial policies of the operation with an interest in gaining economic benefits from its activities. Financial control usually exists if the entity has the right to the majority of the benefits of the operation; however, these rights are conveyed. An entity has financial control over an operation if the operation is considered a group company or subsidiary for the purpose of financial consolidation, i.e. if the operation is fully consolidated in financial accounts.

Each consolidation approach – equity share, operational control and financial control – has advantages and disadvantages. The operational and financial control approaches may best facilitate performance tracking of GHG management policies and be most compatible with the majority of regulatory programmes. However, these may not fully reflect the financial risks and opportunities associated with climate change, compromising financial risk management. On the other hand, the equity share approach best facilitates financial risk management by reflecting the full financial risks and opportunities associated with climate change, but may be less effective at tracking the operational performance of GHG management policies.

The company decides which approach to use to define its organizational boundaries for carbon footprint measurement or GHG inventory. The consultant helping the company in this exercise or the staff in the company preparing the inventory should consult with the company's accounting or legal staff and/or management to understand company's legal structure, each operation's financial accounting category, its financial and operating control policies, etc. Table 2.3 lists out all the financial accounting categories and the organizational boundaries based on equity share and financial control approaches.

Table 2.3 Reporting based on equity share versus financial control

Accounting category	Financial accounting definition	Approach	
		Equity share	Financial control
Group companies/subsidiaries	The parent company has the ability to direct the financial and operating policies of the company with a view to gaining economic benefits from its activities. Normally, this category also includes incorporated and non-incorporated joint ventures and partnerships over which the parent company has financial control. Group companies/subsidiaries are fully consolidated, which implies that 100 percent of the subsidiary's income, expenses, assets and liabilities are taken into the parent company's profit and loss account and balance sheet, respectively. Where the parent's interest does not equal 100 percent, the consolidated profit and loss account and balance sheet shows a deduction for the profits and net assets belonging to minority owners	Equity share of GHG emissions	100% of GHG emissions
Associated/affiliated companies	The parent company has significant influence over the operating and financial policies of the company, but does not have financial control. Normally, this category also includes incorporated and non-incorporated joint ventures and partnerships over which the parent company has significant influence, but not financial control. Financial accounting applies the equity share method to associated/affiliated companies, which recognizes the parent company's share of the associate's profits and net assets	Equity share of GHG emissions	0% of GHG emissions
Non-incorporated joint ventures/partnerships/operations where partners have joint financial control	Joint ventures/partnerships/operations are proportionally consolidated, i.e. each partner accounts for their proportionate interest of the joint venture's income, expenses, assets and liabilities	Equity share of GHG emissions	Equity share of GHG emissions
Fixed asset investments	The parent company has neither significant influence nor financial control. This category also includes incorporated and non-incorporated joint ventures and partnerships over which the parent company has neither significant influence nor financial control. Financial accounting applies the cost/dividend method to fixed asset investments. This implies that only dividends received are recognized as income, and the investment is carried as cost	0% of GHG emissions	0% of GHG emissions
Franchises	The franchiser will not have equity rights or control over the franchise. Therefore, franchises should not be included in consolidation of GHG emission data. However, if the franchiser does have equity rights or operational/financial control, then the same rules for consolidation under the equity or control approaches apply	Equity share of GHG emissions	100% of GHG emissions

Example #1 (Adopted from The Climate Registry (2008))

Given:

Company A has 60 percent ownership and full control of Facility #1 under both the financial and operational control criteria. Company B has 40 percent ownership of the facility and does not have control.

Question:

As the reporting entity, shall Company A or Company B report the emission from Facility #1?

Answer:

Under the equity share approach, Companies A and B would report 60 percent and 40 percent of GHG emissions of Facility #1, respectively, based on their share of ownership. Under either criterion for control, Company A would report 100 percent of GHG emissions for Facility #1, while Company B would report none. 0% means Facility #1 is not within the reporting entity's organizational boundary, but the reporting entity, i.e. Company B in this case, still has the option to report it.

Reporting entity	Ownership of Facility #1 (%)	Equity share approach (%)	Financial control approach (%)	Operational control approach (%)
Company A	60	60	100	100
Company B	40	40	0	0

Example #2 (Adopted from The Climate Registry (2008))

Given:

Company A and Company B each has 50 percent ownership of Facility #1. Company B has the authority to implement its operational and HSE policies, but all significant capital decisions require approval of both Company A and Company B since they have joint financial control.

Question:

As the reporting entity, shall Company A or Company B report the emission from Facility #1?

Answer:

Each reports 50 percent of GHG emissions under the financial control and equity share approaches. Under the operational control approach, Company B reports 100 percent of the facility's emissions, while Company A reports none.

Reporting entity	Ownership of Facility #1 (%)	Equity share approach (%)	Financial control approach (%)	Operational control approach (%)
Company A	50	50	50	0
Company B	50	50	50	100

Example #3: Landlord and Tenant

Hysan Place is a 40-storey retail/office building in Causeway Bay, Hong Kong. Developed by Hysan Development Company Limited, the building gets numerous green building awards and becomes the landmark of Causeway Bay. American fashion brand Hollister California opened its first store in Hong Kong Island in a 20,000 sq. ft. space in its shopping mall. KPMG is the first office tenant.

Given:

KPMG has signed a 9-year lease for the 20th to the 25th floors of the building, taking up an area of approximately 80,000 sq. ft. in Hysan Place, under an operating lease contract.

Question:

Shall the landlord Hysan Development Co. Ltd. and/or the tenant KPMG report the emission from leased office, i.e. 20th to 25th floors?

Answer:

Operating lease enables the lessee to operate an asset, like a building or vehicle, but does not give the lessee any of the risks or rewards of owning the asset. Hence, reporting on an equity share or a financial control approach, as the building owner or the landlord Hysan Development Co. Ltd. owns the building, it should report the emission from the building, which includes the leased office, while as the tenant KPMG does not own (or have a financial interest in) the office building, it would not be required to report emissions associated with the office. Although the tenant is not required to report emissions associated with the leased office when it uses the equity share or financial control approaches, it may opt to report these emissions at their wills.

Using the operational control approach, the tenant KPMG must include all the emissions resulting from leased office, i.e. the 20th to the 25th floors, because it has effective operational control over the space and all of the emission sources within the space. For the building owner or the landlord, it would not be required to report the emissions from the leased space, since effective control over the building's emissions passes to the tenant under an operating lease. The landlord in this case should report the emissions from communal areas, such as the corridors, podium, lift lobbies, car park, sky gardens, etc.

Reporting entity	Equity share or financial control approach	Operational control approach
Landlord	Must report emissions from leased asset	May opt to report emissions from leased asset, but must report emissions from communal areas
Tenant	May opt to report emissions from leased asset	Must report emissions from leased asset

In summary, if the company has an asset such as a leased office or a rented vehicle under an operational lease, the company is required to report the emission if the company is using the operational control approach.

Example #4 (Adopted from The Climate Registry (2008))
Given:
   Alpha, Inc. has five wholly owned or joint operations: Beta, Gamma, Delta, Pi and Omega. The following table outlines the organizational structure of Alpha, Inc.

Question:
   Define the percent of emissions from each of its sub-entities that is included in the parent company's total entity-wide emissions using equity share, operational control and financial control.

Answer:
   As shown in the table below.

Wholly owned and joint operations of Alpha, Inc.	Legal structure and partners	Economic interest held by Alpha, Inc. (%)	Control of operating policies	Treatment in Alpha, Inc.'s financial accounts	Equity share approach (%)	Operational control approach (%)	Financial control approach (%)
Beta	Incorporated company	100	Alpha	Wholly owned subsidiary	100	100	100
Gamma	Incorporated company	40	Alpha	Subsidiary	40	100	100
Delta	Non-incorporated joint venture; partners have joint financial control; other partner is Epsilon	50 by Beta	Epsilon	Via Beta	50 (50% × 100%)	0	50
Pi	Subsidiary of Gamma	75 by Gamma	Gamma	Via Gamma	30% (75% × 40%)	100	100
Omega	Incorporated joint venture; other partner is Lambda	56	Lambda	Subsidiary	56	0	100

2.3.2   Operational Boundary Setting

Operational boundaries determine the direct and indirect emissions associated with operations owned or controlled by the reporting company. Defining the operational boundaries includes to identify emissions associated with its operations, categorize them as direct and indirect emissions and choose the scope of accounting and reporting for indirect emissions.

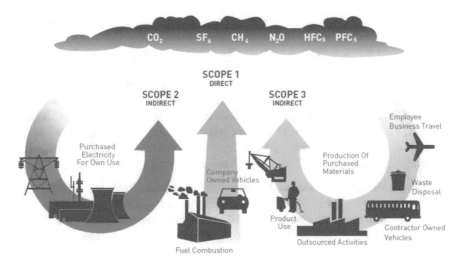

Fig. 2.2   Emissions and scopes. (WBCSD/WRI 2010)

Direct GHG emissions are emissions from sources that are owned or controlled by the company. Indirect GHG emissions are emissions that are a consequence of the activities of the company but occur at sources owned or controlled by another company. What is classified as direct and indirect emissions is dependent on the consolidation approach (equity share or control) selected for setting the organizational boundary. For instance, in Example #3, using operational control approach, the leased office is not in the landlord's organizational boundaries. But the landlord could still report the emission from the leased office as an indirect emission for the landlord.

Based on GHG Protocol (WBCSD/WRI 2010), emissions can be categorized into three scopes, as illustrated in Fig. 2.2.

- Scope 1: All direct GHG emissions (with the exception of direct CO_2 emissions from biomass combustion).
- Scope 2: Indirect GHG emissions associated with the consumption of purchased or acquired electricity, steam, heating or cooling.
- Scope 3: All other indirect emissions not covered in Scope 2, such as upstream and downstream emissions, emissions resulting from the extraction and production of purchased materials and fuels, transport-related activities in vehicles not owned or controlled by the reporting entity (e.g. employee commuting and business travel), use of sold products and services, outsourced activities, recycling of used products, waste disposal, etc.

Scope 1: Direct Emissions
Scope 1 emissions are direct emissions from sources that a company owns or controls and must be reported by the company. The direct emissions are mainly from:

- Stationary combustion of onsite generation of electricity, heat or steam

- Physical or chemical processing, such as the methane gas from the anaerobic digestion, carbon dioxide from cement plant, etc.
- Mobile combustion of fuels used for transportation of materials, products' waste and employees by company's own vehicles
- Fugitive emissions unintentionally released from the production, processing, transmission, storage and use of fuels and other substances that do not pass through a stack, chimney, vent, exhaust pipe or other functionally equivalent opening (such as releases of sulphur hexafluoride from electrical equipment; hydrofluorocarbon releases during the use of refrigeration and air conditioning equipment; and methane leakage from natural gas transport)

The combustion of biomass and biomass-based fuels (such as wood, timber waste, landfill gas, biofuels, etc.) emits GHGs directly. Unlike other fuels, CO_2 emissions from biomass combustion should be reported separately from the scopes, as required by the IPCC Guidelines for National Greenhouse Gas Inventories (IPCC 2006).

Scope 2: Electricity Indirect Emissions
Scope 2 emissions are the indirect emissions that occur when the reporting entity purchases and consumes electricity, heat or steam (generated at a source not owned or controlled by the reporting entity). Scope 2 emissions are a special category of indirect emissions, because purchased electricity represents one of the largest sources of GHG emissions and the most significant opportunity to reduce these emissions. Scope 2 emissions must be reported.

Scope 3: Other Indirect Emissions
Scope 3 includes all other indirect greenhouse gas emissions. It is optional to report Scope 3 emissions, which can be emitted from:

- Extraction and production of purchased materials and fuels
- Upstream transport-related activities, such as transport of raw materials and fuels
- Staff commuting
- Business travels
- Downstream transport-related activities, such as transport of products and waste
- Electricity-related activities not included in Scope 2
- Leased assets, franchise and outsourced activities
- Use of sold products and services
- Waste disposal

> **Example #1**
> This example is from GHG Protocol Corporate Accounting and Reporting Standard, and for further and detailed information on how to account for emissions from electricity-related activities, refer to the WRI/WBCSD GHG Protocol, Corporate Accounting and Reporting Standard (Revised Edition), Chap. 4 and Appendix A (WBCSD/WRI 2010).

Given:

As shown in Fig. 2.3, Company A is an independent power generator that owns a power generation plant. The power plant produces 100 MWh of electricity and releases 20 tonnes of emissions per year. Company B is an electricity trader and has a supply contract with Company A to purchase all its electricity. Company B resells the purchased electricity (100 MWh) to Company C, a utility company that owns/controls the Transmission and Distribution (T&D) system. Company C consumes 5 MWh of electricity in its T&D system and sells the remaining 95 MWh to Company D. Company D is an end-user who consumes the purchased electricity (95 MWh) in its own operations.

Question:

Categorize the emissions from electricity-related activities for different companies.

Answer:

Twenty tonnes of emissions are from the onsite power generation, so it is the direct emission under Scope 1 for Company A.

Company B, as an electricity trader, buys and sells electricity. Electricity is a product of Company B. The embodied carbon emission, i.e. 20 tonnes in this case, is the Scope 3 emission for Company B.

There are two emissions for Company C: 19 tonnes from the generation of 95 MWh purchased electricity that is sold to end-user D and 1 tonne from 5 MWh purchased electricity that it consumes in its T&D. Since Company C is consuming only the 5 MWh associated with its T&D system losses, only the emissions resulting from the generation of this 5 MWh qualify as Scope 2 emissions for Company C. Since Company C does not consume the remaining 95 MWh but rather resells this power, the emissions associated with the 95 MWh represent Scope 3 emissions for Company C.

Emissions from 95 MWh electricity purchased and consumed by Company D is the Scope 2 emission for end-user D, while emissions associated with upstream T&D losses is the Scope 3 for end-user D.

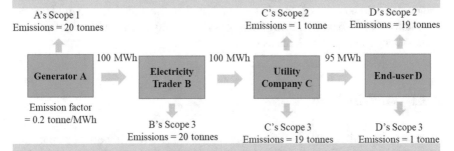

Fig. 2.3 GHG accounting from the sale and purchase of electricity

Example #2

Figure 2.4 summarizes the requirements in Hong Kong, based on the local guidelines (EMSD/EPD 2010).

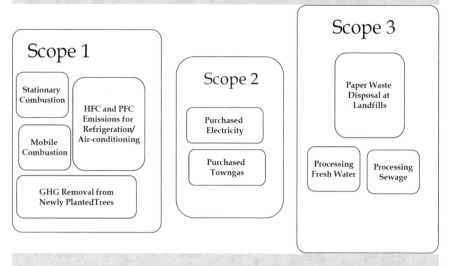

Fig. 2.4 Operational boundaries required by Hong Kong's guidelines. (EMSD/EPD 2010)

2.4 Corporate Carbon Audit: Quantifying the Emissions

After defining the organizational and operational boundaries for the reporting company, the next step is to calculate carbon footprint for each emission source. The most commonly used quantification method is the calculation by using activity data multiple the relevant emission conversion factor. Emission data are then summed up under each scope for a specific operation. For corporate carbon auditing, activity data would be obtained from the reporting company, and the emission conversion factors would be found from the local guidelines. If a specific emission factor cannot be obtained from the local guidebook, IPCC Guidelines and/or other national standards could be referred. In this section, it will be illustrated how to quantify the emissions, mainly based on Hong Kong guidelines, and take reference to IPCC and some other countries' standards as well.

2.4.1 Activity Data

To collect the activity data of the business operation is a very important step in carbon measurement. Table 2.4 lists out some activity data and their possible data sources. Carbon or corporate sustainability consultants should prepare the data collection template for their clients to collect the activity data for carbon auditing. Corporate sustainability personnel also needs to understand where and how they could get the activity data internally before conducting the carbon inventory.

Table 2.4 Activity data and its data source

Emissions	Activity data	Documents
Direct emission from stationary combustion	Fuel consumption	Monthly utility bills, fuel purchase records, inventory of stationary combustion facilities, if any
Direct emission from mobile combustion	Fuel consumption	Gas station card data, fuel purchased records, vehicle mileage data
Direct emission from process	Inputs and outputs	Manufacturing – raw material inputs, production output, chemical reaction, biological process, etc.
Direct emission from refrigeration	Mass of refrigerant leaked	Contractor refill records, refrigerant purchased records
Indirect emission from purchased electricity	Electricity consumption	Electricity bills
Indirect emission from Towngas	Towngas consumption	Towngas bills
Indirect emission from freshwater usage	Water consumption	Water bills
Indirect emission from sewage discharge	Sewage discharge amount	Water bills
Indirect emission from waste disposal	Waste disposed amount	Waste collector records, waste recycling records
Indirect emission from air travels	Flight type and distance	Business trips records, air tickets

2.4.2 Quantifying Scope 1 Emissions

Example #1: Stationary Combustion

Given:

Diesel generator is a main polluter and emission source at a construction site. Assume 1,000 l diesel fuels have been consumed during last month at a construction site in Hong Kong.

Question:

What is the carbon footprint from this stationary combustion?

Answer:

Activity data, AD	Fuel type	CO_2 emission factor, EF_{CO2}[a]	CH_4 emission factor, EF_{CH4}[a]	N_2O emission factor, EF_{N2O}[a]
1,000 litres	Diesel oil	2.614 kg/litre	0.0239 g/litre	0.0074 g/litre
–	LPG	3.017 kg/kg	0.020 g/kg	0.000 g/kg
–	Towngas	2.549 kg/unit	0.0446 g/unit	0.0099 g/unit

[a]The emission factors are from Hong Kong's guidelines (EMSD/EPD 2010)

$$\because CF = AD \times EF_{CO_2} + AD \times EF_{CH_4} \times GWP_{CH_4} + AD \times EF_{N_2O} \times GWP_{N_2O}$$
$$= 1,000 \times 2.614 + 1,000 \times 0.00239 / 1,000 \times 21 + 1,000 \times 0.0074 / 1,000 \times 310$$
$$= 2.616 \, tonnes \, CO_2e$$

$\therefore$ Carbon footprint of the activity – stationary combustion of diesel fuel from onsite generator is 2.616 tonnes of CO_2e.

Example #2: Mobile Combustion

Given:

Company X in Hong Kong owns a passenger car and consumed 1,000 litres of gasoline during the last fiscal year.

Question:

How much carbon has been emitted from this vehicle during the past year?

Answer:

Activity data, AD	Fuel type	CO_2 emission factor, EF_{CO2}[a]	CH_4 emission factor, EF_{CH4}[a]	N_2O emission factor, EF_{N2O}[a]
–	Diesel oil	2.614 kg/litre	0.072 g/litre	0.110 g/litre
1,000 litres	Unleaded petrol	2.36 kg/litre	0.253 g/litre	1.105 g/litre

[a]The emission factors are from Hong Kong's guidelines (EMSD/EPD 2010)

$$\because CF = AD \times EF_{CO_2} + AD \times EF_{CH_4} \times GWP_{CH_4} + AD \times EF_{N_2O} \times GWP_{N_2O}$$
$$= 1,000 \times 2.36 + 1,000 \times 0.253 / 1,000 \times 21 + 1,000 \times 1.105 / 1,000 \times 310$$
$$= 2.708 \, tonnes \, CO_2e$$

$\therefore$ Carbon footprint of the activity – mobile combustion of unleaded petrol from the company-owned passenger car of last year is 2.708 tonnes of CO_2e.

Example #3: Emissions from Biofuels

Biofuels are treated as cleaner and lower emission fuels. Bioethanol is an alcohol made by fermenting the sugar components of plant materials, mostly from sugar and starch crops. Biodiesel is made from vegetable oils, animal fats or recycled greases. With the more widely acceptable use of biofuels for vehicles in business, corporate sustainability executives or carbon consultants are usually the ones who are approached and asked how much carbon is reduced by using biofuels, replacing the diesel or gasoline. However, there are no international standards on calculating carbon emission from the combustion of biofuels. In addition, lots of local governments haven't published their biofuel emission factor. In addition, it has been challenged whether biofuels are green alternatives to fossil fuels, as biofuels need a lot of land which reduces the carbon sink, the manufacturing process of biofuels is not efficient, and the combustions of biofuels also emit GHGs (Steer and Hanson 2015).

While it is difficult to calculate the emissions from biofuels, I would like to discuss come concepts and introduce some treatment of biofuels in carbon audit in this section. Firstly, this book is talking about the anthropogenic GHG emission, and due to its biogenic nature, emission from burning of biofuels cannot be reported under Scope 1, but should be reported separately. Secondly, we still need to report the emissions from non-biogenic portion of the biofuels. The following example will illustrate the calculation and reporting method for biofuels.

Given:

A US agriculture company used 1,000 gallons of B20 for its tractor for the last farming season.

Question:

What is CO_2 emission of this activity, and how much CO_2 has been reduced compared with use of the same amount of diesel for this tractor?

Answer:

$$\therefore \text{B20} = 20\% \times \text{B100} + 80\% \times \text{diesel}.$$

$$\therefore \text{Direct } CO_2 \text{emission from diesel} = 1,000 \times 80\% \times 10.15 = 8.12 \text{ tonnes } CO_2$$

$$\text{Reduced } CO_2 \text{emission} = 1,000 \times 20\% \times 10.15 = 2.03 \text{ tonnes } CO_2.$$

$$CO_2 \text{emission from biogenic source} = 1,000 \times 20\% \times 9.46 = 1.89 \text{ tonnes } CO_2.$$

The above emission factors for diesel and B100 are from US guidelines (The Climate Registry 2008).

Example #4: Process Emissions

Direct emissions from anaerobic wastewater plant:

$$CH_4 \text{emission} = (OC - S) \times EF - R.$$

where:

OC = BOD or COD enters anaerobic treatment system
S = Organic content removed in the sludge
EF = Emission factor
R = Methane recovery, capture or flaring onsite

The calculation methodology and emission factors can be found from IPCC 2006, Guidelines for National Greenhouse Gas Inventories, Volume 5, Wastewater (IPCC 2006).

Example #5: Fugitive Emissions

Given:

You commissioned a contractor to refill the refrigerants for your facility and got a summary table from your contractor as below:

Refrigerant	GWP	Amount, kg
CFC-13	14,400	100
HCFC-22 (R22)	1,500	100
HFC-23	11,700	100

Question:

What is the carbon footprint of fugitive emissions of the refrigerants?

Answer:

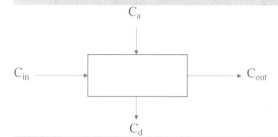

C_{in} is the refrigerant at the beginning of the inventory period;

C_{out} is the refrigerant in the end of the inventory period;

C_a is the refrigerant added by the contractor during the inventory period;

C_d is the refrigerant disposed

Fig. 2.5 Mass balance of the refrigerant

Activity data of fugitive emission from the refrigerants is the released weight of the refrigerants, $C_{released}$. Figure 2.5 shows the mass balance of the refrigerant.

$$C_{released} = C_{in} + C_a - C_{out} - C_d.$$

In majority of the cases, like this one, how much refrigerant at the beginning of the reporting period and how much left in the end of the reporting period are unknown. In addition, it is assumed there is no disposal during the reporting period. Therefore, $C_{released} = C_a$.

Refrigerant	Activity data, Kg	Emission factor, GWP	Carbon footprint, tonnes CO_2e
CFC-13	100	14,400	1,440
HCFC-22 (R22)	100	1,500	150
HFC-23	100	11,700	1,170

It should be noted here that carbon footprint measurement only covers the Kyoto gases, as we mentioned in Sect. 1.7. CFCs and HCFCs should not be reported under Scope 1, but be reported separately. CFCs and HCFCs are commonly used types of refrigerants, and some also have very high global warming potentials. The reason of not reporting them is that as high ozone-depleting gases, they are well covered by Montreal Protocol, which is the most successful global treaty in combating ozone layer depletion.

The direction emission from fugitive emission of refrigerant in this example is then 1,170 tonnes.

Example #6: GHG Removal

Trees and plants could remove carbon dioxide from the atmosphere during the photosynthesis process. Trees and plants are called carbon sink. Although for GHG protocol corporate accounting, there is no method on how to account GHG removal from natural sequestration, some local government would publish their own method or emission factor for sequestration. In this case, GHG removal should be reported under Scope 1.

Given:

Company A planted 2,000 trees in Hong Kong in its last reporting year.

Question:

How much carbon has been removed for Company A last year?

Answer:

Based on Hong Kong's guidelines (EMSD/EPD 2010).

CO_2 removed by trees in 1 year = net number of additional trees planted since the concerned building is constructed × removal factor (estimated at 23 kg/tree∗).

∗The figure is applicable to all trees commonly found in Hong Kong which are able to reach to at least 5 metres in height.

So, is the answer simply

23 × 2,000/1,000 = 46 tonnes carbon removal?

It should be noted that carbon removal is the GHG sequestration that happened within the organizational boundary of the reporting entity. It is different from carbon offsetting, which is happening outside of the organizational boundaries. Hence, in this example, we should ask if the trees have been planted in Company A's organizational boundary or in the country park, owned by Hong Kong government. If it's the former, the answer is 46 tonnes of carbon removal in Scope 1.

2.4.3   Quantifying Scope 2 Emissions

Scope 2 emissions are indirect emissions from purchased electricity, heat and gas that are produced outside of the reporting company's boundaries.

$$CF_{Scope\,2} = \Sum\left(AD_i \times EF_i\right)$$

AD_i is the electricity or gas consumption data
EF_i is the relevant emission factor from that utility

The emission factors for Scope 2 emissions are normally released by the utility companies, such as the power plants. Local governments will publish the emission factors, but most updated emission factors could be found from electricity and gas companies' annual report and/or their sustainability reports. For example, Table 2.5 lists out the emission factors for the two power companies (i.e. China Light Power, CLP, and Hong Kong Electric, HEC) from the Hong Kong government released guidebook for carbon auditing (EMSD/EPD 2010), which was updated in 2010, and the most updated emission factors (i.e. from 2009 till 2017) are summarized in Table 2.6.

It is the same to Towngas in Hong Kong, which is summarized in Table 2.7. One major job for carbon consultants and corporate sustainability practitioners, who need to conduct the carbon footprint measurement, is to collect and compile the emission factors, especially the reporting companies in those countries where there are no published utility emission factors or the data haven't been updated very frequently.

Table 2.5 Emission factors for Hong Kong's power companies (in kg CO_2e/kWh) (EMSD/EPD 2010)

Power company	2002	2003	2004	2005	2006	2007	2008
CLP	0.48	0.56	0.53	0.52	0.53	0.57	0.54
HEC	0.96	0.98	0.98	0.92	0.91	0.83	0.84

Table 2.6 Emission factors for Hong Kong's power companies (in kg CO_2e/kWh)

Power company	2009	2010	2011	2012	2013	2014	2015	2016	2017
CLP[a]	0.56	0.54	0.59	0.58	0.63	0.64	0.54	0.54	0.51
HEC[b]	0.81[a]	0.79[a]	0.79	0.79	0.78	0.79	0.78	0.79	0.79

[a]From CLP group's sustainability reports
[b]From HEC's sustainability reports

Table 2.7 Emission factors for gas in Hong Kong (in kgCO_2e/unit purchased Towngas[a])

Year	2005	2006	2007	2008	2009	2010	2011	2012	2013	2014	2015	2016	2017
EF	0.735	0.693	0.592	0.593	0.628	0.620	0.618	0.610	0.620	0.600	0.605	0.599	0.592

[a]Emission factors for Towngas for 2005–2008 are from EMSD/EPD (2010), while those for 2009 till 2017 are from Towngas' sustainability reports

2.4.4 Quantifying Scope 3 Emissions (Optional)

Indirect emissions are those emissions not within the reporting entity's organizational boundaries, hence no need to be reported. But since Scope 2 emission is of paramount importance to the reporting organization, one has to report Scope 2 emission, while Scope 3 emissions are all the other indirect emissions, which is optional to be reported. In this section, I just give two examples to illustrate how to calculate Scope 3 emissions.

Example #1: Indirect Emissions from Water Consumption and Sewage Discharge

Given:

Restaurant A in Central, Hong Kong, consumed 10,000 m^3 tap water in 2016.

Ask:

Calculate the indirect emissions from this activity.

Answer:

This calculation takes into account the indirect GHG emissions due to electricity and energy used for processing fresh water and treating wastewater at water/sewage treatment plants. The water consumption data from water bills serves the activity data for both water use and sewage discharge. However, it should be noted that while we assume 100% of the fresh water consumed will enter the sewage system in common commercial and residential buildings, we assume 70% of the fresh water consumed will enter the sewage system for restaurants and catering services (EMSD/EPD 2010).

$$CF_{water} = AD_{water} \times EF_{water} = 10,000 \times 0.402^* / 1,000 = 4.02 \text{ tonnes CO}_2\text{e}.$$

$$CF_{sewage} = AD_{sewage} \times EF_{sewage} = 10,000 \times 70\% \times 0.190^{**} / 1,000 = 1.33 \text{ tonnes CO}_2\text{e}.$$

*Emission factor from Hong Kong's Water Service Department's annual report.

**Emission factor from Hong Kong's Drainage Service Department's sustainability report.

Example #2: Indirect Emissions from Air Travel
Table 2.8 shows how to calculate indirect emissions from business trips. Based on UK's standards (DEFRA 2018), emission from air travel depends on the type of the flight, different cabin classes and the distance of travelling.

Table 2.8 Emission factors for business air trips (DEFRA 2018)

Flight	Class	kg CO$_2$e/ passenger/km
Domestic, to/from the UK	Average passenger	0.29832
Short-haul, to/from the UK	Average passenger	0.16236
	Economy class	0.1597
	Business class	0.23955
Long-haul, to/from the UK	Average passenger	0.21256
	Economy class	0.16279
	Premium economy class	0.26046
	Business class	0.47208
	First class	0.65115
International, to/from the non-UK	Average passenger	0.18277
	Economy class	0.139965
	Premium economy class	0.22395
	Business class	0.4059
	First class	0.55987

2.5   Tracking Emissions

Tracking GHG emissions over time is the same as monitoring the sustainability performance of the reporting company. It enables the reporting company to measure its emissions against the targets to understand where it is and how it has performed, to report its GHG reductions and to manage the potential risks and opportunities. In this section, we will discuss how to choose a base year and whether the baseline should be recalculated and if yes, how.

2.5.1   Establishing Baseline

A "base year" is a benchmark against which an entity's emissions are compared over time. The reporting company's base-year emission is called baseline. How to choose a base year and establish its baseline?

- It can be the earliest reporting year the company submits a complete emission report – a report that fulfils the reporting standard that the company chooses.

- It can also be a historical year when the company submits complete data or all subsequent years.
- The base year could be a calendar year or a fiscal year.
- Business growth should be taken into account when choosing the base year. For instance, most companies in Hong Kong would not choose 2003 as their base year, as the outbreak of the deadly infectious disease SARS in Hong Kong happened in 2013 followed by a sharp drop in GDP. This unexpected valley point of business growth would not provide a good benchmarking point for emissions, as we know that anthropogenic GHG emissions are highly dependent on the consumptions and activities.

2.5.2 Updating Baseline

Why need to update baseline?

- Companies often undergo significant structural changes such as acquisitions, divestment and mergers.
- Some emission scopes can be changed over time, e.g. from Scope 1 to Scope 3, if the activity is outsourced.
- Companies may also make different kinds of mistakes during GHG inventory preparation and carbon auditing.

Setting a base year allows to scale structural changes to their entity back to a benchmarked emission profile. Adjustments to base-year emission – baseline – are generally made to reflect organizational changes such as mergers, acquisitions or divestments.

How to update? The reporting company should develop a baseline recalculation policy and define "significant threshold"– a qualitative and/or quantitative criterion used to define any significant change to the data, boundaries, methods or other relevant factors. For instance, 5% threshold is used for the US Climate Registry (2008), while 10% for California Climate Action Registry (2009). The reporting company should update its baseline when the changes are beyond the defined significant threshold. The following examples illustrate the different cases on how to do it.

Example #1: Acquisition and Mergers
Company A's GHG emission base year is 2010 and the baseline is 100 tonnes CO_2e. In year 2011, Company A acquired Company B, with emission in 2010 of 3 tonnes CO_2e. If we set the significant threshold as 5%, the acquisition of Company B was not significant (i.e. 3% <5%), so that in 2011 the baseline for Company A remained unchanged, that is, 100 tonnes CO_2e.

Example #1 (continued)

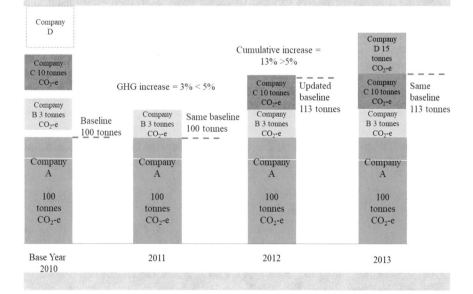

If in 2011, Company A merged with Company C and Company C's emission in 2010 was 10 tonnes, which was more than 5% of the Company A's baseline, then the baseline of Company A must be updated in 2011 to reflect the acquisition. The updated baseline is 110 tonnes CO_2e.

Example #2: Cumulative Acquisitions
Company A acquires Company B, C and D over 3 years.

In 2011, the baseline remained unchanged, as the acquisition of Company B represents less than 5% increase. In 2012, acquisition of Company C amounts to a 13% cumulative increase in base-year emission, which is significant and the baseline is adjusted to 113 tonnes CO_2e.

In 2013, Company A acquired Company D, which came into existence in 2011. In this acquisition case, the baseline of Company A should not be updated, because the acquired Company D did not exist in the base year (or, in other words, the base-year emissions of Company D were zero). Hence in 2013, the baseline of Company A remained unchanged, that is, 113 tonnes CO_2e.

Example #3: Cumulative Divestiture
Company A divests three divisions X, Y and Z over 3 years. Each of these divisions accounts for 4% of Company A's GHG emissions.

Because the cumulative effect of these divestitures reduced Company A's emissions by more than 5% in 2012, baseline should be adjusted in 2011 to 92 tonnes CO_2e.

Example #4: Organic Growth and Decline

Company X consists of two business units (A and B). In its base year, 2010, each business unit emitted 50 tonnes CO_2e. The baseline for Company X is 100 tonnes CO_2e. In 2011, both business units increased productivity and produced more products, leading to an increase in both emissions to 60 tonnes CO_2e per business unit. The total emission for Company X in 2011 was 120 tonnes CO_2e. The baseline could not be updated in this case, as the increase in emission was due to "organic growth", which includes business expansion, increase in output, increase in customer base, new products development, etc.

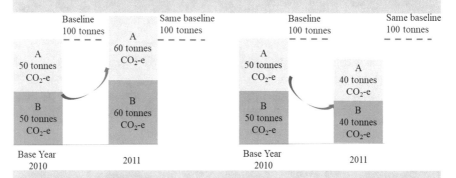

The same principle applied to business organic declined. In this case, in 2011 both units reduced production, resulting in a decrease in emission, i.e. 10 tonnes CO_2e for each unit. The baseline for Company X in 2011 could not be adjusted.

Example #5: Outsourcing and Insourcing

Company A owned and operated shuttlebus services in base year 2010, but contracted out this service to a public bus company in 2011. In 2010, the baseline for Company A was 100 tonnes, and 6 tonnes emission from owned shuttlebus was Scope 1 emission. In 2011, this emission became emission from outsourced services under Scope 3.

Company A has an option to choose to report or not to report Scope 3 emission. If in 2011, Company A decided not to report 6 tonnes emission under Scope 3, then since 6% is more than the threshold, Company A must adjust its baseline in 2011 and the updated baseline was 94 tonnes CO_2e. But if Company A decided to report this emission under Scope 3, then Company A could not update its baseline in 2011.

Owned shuttlebus 6 tonnes CO_2-e	Outsourced shuttlebus 6 tonnes CO_2-e	Outsourced shuttlebus 6 tonnes CO_2-e	Owned shuttlebus 6 tonnes CO_2-e
Rest Emission of Company A	Rest Emission of Company A	Rest Emission of Company A	Rest Emission of Company A
94 tonnes CO_2-e	94 tonnes CO_2-e	94 tonnes CO_2-e	94 tonnes CO_2-e
Base Year 2010	2011	Base Year 2010	2011

In the insourcing case, if in base year 2010, Company A's baseline was 94 tonnes, i.e. not reporting 6 tonnes emission under Scope 3, then in 2011, when Company A insourced this service, it should adjust its baseline to 100 tonnes. If in base year 2010, Company A reported 6 tonnes emission under Scope 3 and its baseline was 100 tonnes, then in 2011, Company A could not adjust its baseline.

From this example, it is noted that the baseline could not be updated for outsourcing/insourcing if emissions are already reported under Scope 3. In addition, if emissions are not reported under Scope 3, the baseline should be updated if the change is significant, which avoids the reporting company to use outsourcing as a reduction measure.

Case: Coca-Cola European Partners

Carbon footprint baseline recalculation policy for Coca-Cola European Partners shows a good example of how to establish such a policy when conducting carbon management for the company, which is also a good summary of what we have discussed in Sect. 2.5.

Base-Year Changes

2007 will be our base year. For purposes of accurately tracking progress towards our goals, we will adjust our base-year emission inventory for significant qualitative or quantitative structural changes or methodology changes (as defined below).

Base-year recalculation will be reported with the first full reporting year after the change occurs (e.g. change occurs in 2010, first full year of data in 2011, recalculation of base year reported with the 2011 CRS report published in 2012).

When recalculation is triggered, other changes of non-significance will also be added to the base year under the principles of completeness and accuracy.

Structural Changes

Structural changes are acquisitions, divestures or mergers of facilities that existed during our base year.

Where such addition or removal of such facilities would reflect a quantitative change greater than our significance threshold in the base-year inventory, we will endeavour to add or delete as appropriate the emissions associated with that facility from our base year. In sourced or outsourced operations will be treated similarly, unless those operations will be reflected in our inventory as Scope 3.

An exception to the above would be where the acquisitions, divestures or mergers relate to the addition or removal of one or more business units (this generally refers to a country or group of countries like GB); this would reflect a quantitative change that will not need to meet the significance threshold. We will endeavour to add or delete as appropriate the emissions associated with that business unit from our base year.

Methodology Changes

Methodology changes may include updated emission factors, improved data access or updated calculation methods or protocols. Where such methodology changes would reflect a change greater than our significance threshold in the base-year inventory, we will implement the change at a minimum in our base-year inventory and our current year inventory. We may optionally implement the change in all interim year inventories.

Other Changes

We will endeavour to add or delete as appropriate the emissions associated with the discovery of significant errors, or a number of cumulative errors, that are collectively greater than the significance threshold.

Changes to our reporting boundaries, for example, inclusion of previously unreported Scope 3 emissions (i.e. waste). We will endeavour to add or delete as appropriate the emissions associated with that element from our base year regardless of the significance threshold.

Insourcing or outsourcing that shifts significant emissions between Scope 1 and Scope 3 when Scope 3 is not reported will trigger a base-year emission recalculation. Other non-organic growth or decline changes that would compromise the consistency and relevance of the reported GHG emissions may also warrant a recalculation of the base year and may be excluded in our carbon footprint until they are represented in the base year. The reasons why must be documented.

Significance Threshold
CCE will recalculate base-year emissions if a change in organizational structure or data quality exceeds a significance threshold of 5% of base-year emissions. We will review this significance threshold on an annual basis.
   Source: https://cokecce.com/system/file_resources/7/Carbon_Footprint_Baseline_Re-calculation_Policy.pdf

2.6   Carbon Audit Process

The term carbon audit is commonly used for business and consultants. Carbon audit is a means of measuring and reporting GHG emission in carbon footprint by following the GHG Protocol and adhering to international standards such as ISO 14064 and/or local standards. Carbon audit is different from traditional environmental audits, such as compliance audit and environmental management system (EMS) audit, because traditional environmental audits identify the compliance and system gaps and action plans for continuous improvement, while carbon audit measures and reports the performance. There are no so-called gaps or nonconformity issues in carbon audit, but it is highly recommended that the reporting company or the consultants who conduct the carbon audit follow the international or local reporting requirement and identify the feasible reduction solutions and action plans in the carbon audit report, even if not required.

Carbon audit process includes the following steps:

- Statement of work (for proposal)
- Project kick-off meeting
- Site visit
- Data collection
- Calculation
- Reporting

Table 2.9 shows the statement of work I use for my proposal preparation. Basically, it lists out all the tasks in a carbon audit job. It also tells who will be involved – your clients or you as a consultant – and how many man-days are needed for each task. It helps your potential client understand better the carbon audit process and also helps with the total cost and budget allocation.

It should be noted that when you are commissioned by your client to help conduct the carbon audit job, you as a consultant should spend time with your client to understand their business objectives in measuring and reporting their GHG emission, their organization structure, their business operations and activities, etc. You can then define the organizational and operational boundaries for your client and define which standard to use for reporting. You then need to understand your client's current data system, to figure out where you can get the activity data. In most cases, especially when your client doesn't have an online data capture system, you need to

Table 2.9 An example of statement of work

Task	Content	People involved	Days
1.1 Introductory meeting and site visit	Discuss the business need and assess the current situation for developing a carbon management strategy	Client and consultant	
1.2 Setting boundaries	Consult to set organization and operational boundaries – Including emission sources	Client and consultant	
1.3 Defining standards and rules	Define protocols for auditing carbon footprint using primarily GHG guidelines and possible specific requirements from the client	Client and consultant	
1.4 Setting out data gathering requirements	Develop templates and standard protocols for data collection	Consultant	
1.5 Data gathering	Collect data required for the carbon audit	Client	
1.6 Carbon footprint calculation	Calculate total emissions and analyse emission profile of using the data collected	Consultant	
1.7 Carbon reporting	Compile final carbon audit report, with reduction analysis and recommendations	Consultant	
1.8 Documentation and presentation of carbon footprint and recommendations	Completed carbon audit report will be presented along with all the documentation about the development of the audit and suggestion for any future mitigation or reduction plans	Client and consultant	

prepare the data collection template for your client. Your client will identify internal data owners for different activities and collect the necessary data for carbon footprint calculation.

Third-party verification is defined as an independent expert assessment of the accuracy and conformity of a carbon audit report based on reporting requirements. Although verification is not mandated or required, it is recommended that the reporting company could source a third party to verify the accuracy of the scope, boundaries, data source, calculation, etc.

2.7 Carbon Measurement and Reporting Tools

2.7.1 Online Calculator

There are many online carbon calculators developed by national or local government, NGOs and consulting firms. When different calculators have their own features, overall, they are developed for the public awareness and education purpose, to help the users get the carbon footprint data easily and quickly, without a deep understanding on the methodologies we have discussed above. A few examples are listed below:

CarbonCare Asia Ltd.: https://www.carboncareasia.com/eng/Carbon_Solutions/
    Carbon_Calculators.php
Carbon Footprint Ltd.: https://www.carbonfootprint.com/calculator.aspx
Hong Kong Environment Bureau: https://www.carboncalculator.gov.hk/en
Taiwan EPA: https://ecolife.epa.gov.tw/Cooler/English/eng_Calculator.aspx
The Nature Conservancy, USA: https://www.nature.org/en-us/get-involved/how-to-
    help/consider-your-impact/carbon-calculator/
The Resurgence Trust: https://www.resurgence.org/resources/carbon-calculator.
    html
USEPA: https://www3.epa.gov/carbon-footprint-calculator/
WWF, UK: https://footprint.wwf.org.uk/

2.7.2 Carbon Reporting and Disclosure

While carbon calculators are used for public for raising awareness and education, corporate needs more sophisticated tools for carbon reporting. There are existing voluntary climate-related reporting frameworks, including those developed by the CDP (formerly the Carbon Disclosure Project), the Climate Disclosure Standards Board (CDSB), the Global Reporting Initiative (GRI), the International Integrated Reporting Council (IIRC) and the Sustainability Accounting Standards Board (SASB). Some stock exchanges like HKEX (Hong Kong Stock Exchange) and SGX (Singapore Exchange) have mandated the Environmental, Social and Governance (ESG) reporting requirements for their listing companies, where GHG emission is one of the environmental KPIs, which needs to be measured and disclosed annually.

While most companies just follow the voluntary or mandatory framework to disclose their carbon value alone or as an environmental performance parameter on their sustainability report or ESG report, some companies like Puma and Kering have already taken the lead to price the invisible environmental impact and put them into the environmental financial statement, i.e. environmental P&L. The Task Force on Climate-Related Financial Disclosures (TCFD) has called on all public companies to report on climate risks and how they address them in mainstream financial filings. By doing this, it is hoped in the near future, climate-related issues could be viewed as mainstream business and investment considerations and be embedded in the business strategy and decision-making process, which actually should be as soon as possible, as we don't have long time to the tipping point.

More information on the above reporting framework can be found in:

- CDP https://www.cdp.net/en
- CDSB https://www.cdsb.net/
- GRI https://www.globalreporting.org/Pages/default.aspx
- IIRC http://integratedreporting.org/the-iirc-2/
- SASB https://www.sasb.org/

- HKEX ESG Reporting https://www.hkex.com.hk/listing/rules-and-guidance/other-resources/listed-issuers/environmental-social-and-governance?sc_lang=en
- SGX ESG Reporting http://rulebook.sgx.com/net_file_store/new_rulebooks/s/g/SGX_Mainboard_Practice_Note_7.6_July_20_2016.pdf
- Puma https://glasaaward.org/wp-content/uploads/2014/01/EPL080212final.pdf
- Kering http://www.kering.com/sites/default/files/kering_group_2016_epl_results.pdf
- TCFD https://www.fsb-tcfd.org/

2.7.3   Software Solutions

When a company wants to measure and report its GHG emission, the company can commission a consultant to conduct a carbon audit or the company itself can source a software solution, either hosting it on its own server or subscribing a web-based software solution to do the job. A few examples of carbon measurement and reporting software tools are listed below:

- Carbon Footprinting Software by Carbon Trust
- Greenhouse Gas Emissions Software solution by Enablon
- ghgTrack by ADEC Innovations

The main features of software solutions include the following:

- It can adopt different GHG emission accounting and reporting standards and allows a company to report its carbon footprint based on internationally or country-wise recognized greenhouse gas emission standards.
- It can measure and manage emissions with company-wide operations in an intelligent way, especially for complicated organizational structures in different geographical areas.
- It can provide a real-time emission profile. Management team could have quicker decision and adjust their plan, target and programmes more quickly if they can better track the carbon performance. Not like using the Excel file, both data collection and reporting would have a time delay.
- It normally has a user-friendly and efficient data entry. The company hierarchy, emissions profile and the assigned emission source structure the entry of information underlying the carbon footprint. Some software solutions also enable batch data upload by using the provided template, and others also allow for integration to existing transactional system in order to reduce or eliminate manual data entry.
- It can adopt different and flexible reporting format, such as different dashboard with emission, business measures and abatement. It can also report at different levels and for different internal users.

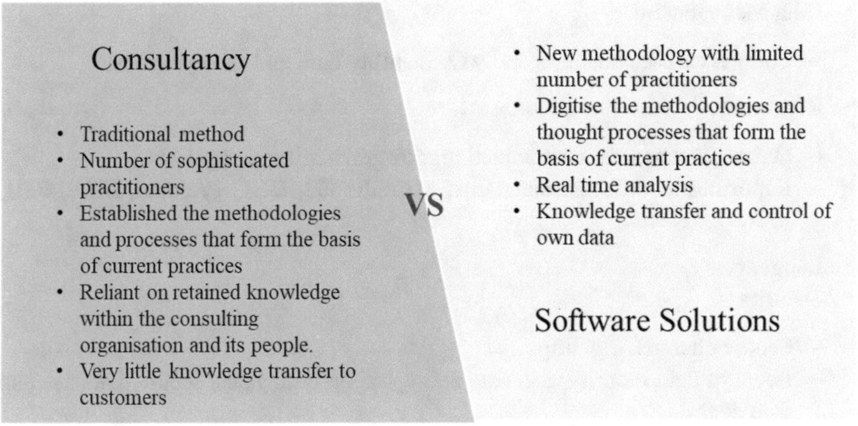

Fig. 2.6 Comparison between traditional consultancy services and software solutions

Figure 2.6 illustrates the difference between using a traditional consultancy service and hosting or subscribing a software solution. Online software solution could track carbon performance in a real-time mode, and carbon measurement knowledge could be transferred to the reporting company that the company could better understand its carbon performance and manage it more effectively and efficiently. But seasoned consultants have broader experience in carbon management; especially in different scenarios, they could offer valuable suggestions and recommendations. As a corporate sustainability practitioner, it is always best to collaborate with good consultants, equip yourselves with sufficient knowledge and use your own tools.

How to Procure a Software Solution or Service?

- Data Capture

  - Manual data entry via customized web forms or preconfigured spreadsheet templates
  - Automated data capture

- Users

  - Administration role: full control of the system
  - Supervisory role: oversee specific areas assigned to his/her subordinates
  - Structure users: data input to assigned areas

- Task Management

  - Prompter/emails to remind data owners to input data regularly
  - Real-time alerts and notifications based on pre-set thresholds of consumption
  - Analytics for targeting, forecasting, variance analysis, benchmarking and ranking

- Data Management

  - Energy, carbon, business and sustainability data and life cycle cost

- Reporting

  - Online dashboards, customized reports and audit trail reports
  - Reporting to different standards, e.g. GRI G3, GHG Protocol, ISO 14001, ISO 50001, etc.

- Languages
- Security

  - Secure channel, e.g. https
  - Level of data security, e.g. data encrypted on both sides, secure database and data centre.

- Others

  - Customized front page and other tailor-made functionalities
  - Local technical support (24/7)

2.8   Product Carbon Footprint

2.8.1   What Is Product Carbon Footprint?

Product carbon footprint (PCF), or carbon footprint of a product, is defined as the sum of GHG emissions and removals in a product system, expressed as CO_2e and based on a life cycle assessment using the single impact category of climate change (ISO 2018b).

Figure 2.7 shows the relations between corporate GHG emission as we have discussed above and the PCF as going to be discussed now. If a manufacturing company produces product A, the company needs to report Scope 1 and 2 emissions, while Scope 3 emission is an option for the company to report or not (WBCSD/WRI

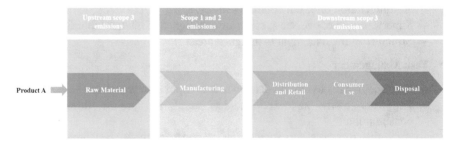

Fig. 2.7   Relation between corporate GHG emission and PCF

2011). Calculating the PCF of product A is basically extending the emission scopes, not only covering the emission from production at manufacturing company site but also covering the emission from Scope 3, from both upstream (i.e. material acquisition, pre-processing of the materials and products, the related transportation, etc.) and downstream (i.e. distribution, storage, retailer, users, end of life, etc.) emissions.

Why company wants to conduct PCF for their products? The possible reasons might be to:

- Provide a comparison among products
- Improve overall environmental performance
- Justify environmental marketing claims
- Develop strategic or tactical goals
- Enhance brand positioning
- Have competitive advantage in "green" purchasing market
- Identify potential cost savings

2.8.2 Life Cycle Assessment (LCA)

Life cycle assessment (LCA) is a tool for the systematic evaluation of the environmental aspects of a product or service system through all stages of its life cycle. LCA provides an adequate instrument for environmental decision support. Reliable LCA performance is crucial to achieve a life cycle economy.

LCA is defined as compilation and evaluation of the inputs, outputs and the potential environmental impacts of a product system throughout its life cycle (ISO 2018b). And PCF LCA is only focusing on a single impact category: climate change.

There are four phases of life cycle assessment:

- Goal and scope definition
- Life cycle inventory analysis
- Impact assessment
- Interpretation and evaluation

ISO has standardized this framework within the series ISO 14040, as summarized in Fig. 2.8. The following standards are normally used for PCF LCA:

- The GHG Protocol Product Life Cycle Accounting and Reporting Standard, WBCSD/WRI, 2011
- ISO 14067: 2018 Greenhouse gases — Carbon footprint of products — Requirements and guidelines for quantification
- PAS 2050: 2011 Specification for the assessment of the life cycle greenhouse gas emissions of goods and services

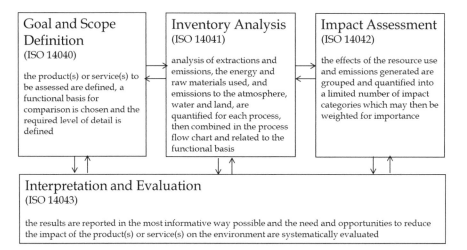

Fig. 2.8 Framework of LCA

2.8.3 Steps of PCF LCA

Step 1: Defining Business Goals

Companies should identify their business goals before conducting PCF project – why do they want to do PCF and what kind of benefits would bring to the companies in doing this?

PCF is basically extending the carbon measurement and GHG emission inventory from the organizational boundaries to its supply chains that cover the product's life cycle. Hence, besides the basic benefits of measurement, such as knowing where you are, how good or bad your carbon performance is, how to track and monitor your performance and where to figure out the potential emission reduction and cost savings, PCF project could bring the following additional benefits to their companies:

- Opportunities and risks

PCF helps the company to understand the opportunities and risks along a product's life cycle. Both opportunities and risks could be resulted from the current and future differences in climate legislation or incentives in different geographic locations for raw materials or manufacturing. Emission reduction, cost saving and better cost forecasting could be better facilitated, and risks could be better managed through PCF LCA project for their products and services.

- Supply chain management

PCF project provides the company a platform to engage its stakeholders along the product's supply chain or value chain, which includes raw material suppliers, logistic companies, distributors, retailers, final users, etc. PCF, in other words, could accelerate the supply chain management for the company, by understanding the compliance issues, the risks and the potential savings and could drive especially both suppliers and users to contribute to lower carbon footprint and climate impact of the product.

- Market demand

PCF is a good parameter for companies to compare its climate impact with its competitors. Whether it is a demand from the customers or not, PCF provides potential competitive advantage and brand differentiation from others, when the company's PCF is better measured and managed.

Step 2: Defining Inventory Scope
Defining inventory scope includes the following steps:

- Identify which GHGs to account for

  - Emissions and removals of six Kyoto gases, both direct and indirect, during the product's life cycle
  - Additional GHGs included shall be listed separately in the inventory report.

- Choose a product

  - Choose the product for study to conduct PCF project

- Define the unit of analysis

  - The unit of analysis is defined as the performance characteristics and services delivered by the product being studied
  - Assessment of the GHG emissions arising from the life cycle of products shall be carried out in a manner that allows the mass of CO_2e to be determined per functional unit for the product
  - Where a product is commonly available on a variable unit size basis, the calculation of GHG emissions shall be proportional to the unit size (e.g. per unit, per kilogram or per litre of products sold or per month or year of a service provided).

Step 3: Setting Boundary
The system boundaries determine which unit processes need to be included in the LCA study. It starts with defining the life cycle stages. As illustrated in Fig. 2.9, there are different approaches to define the overall boundary by using different life cycle stages involved in PCF project:

- Cradle-to-grave: it is the full LCA from raw material extraction or production ("cradle") to disposal of the product after using ("grave"). This is a usual procedure to be applied on commodities.
- Cradle-to-cradle: it is an extended scope of cradle-to-grave assessment, where the end-of-life disposal for the product is a recycling process. This is the most complex one because it includes everything taken into account by the cradle-to-grave process but also the recirculation of what could be considered as waste into the process again by recycling or reusing them as raw materials of another product as the first stage of the process. Cradle-to-cradle is a method used to minimize

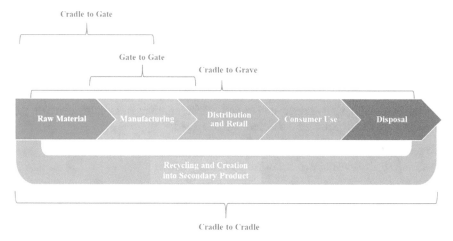

Fig. 2.9 Life cycle stages for the system boundary of LCA

the environmental impact of products by employing sustainable production, operation and disposal practices and aims to incorporate social responsibility into product development as well. It is also called "close-the-loop" or "circular economy", which will be discussed further in Chap. 7 on the solution to resources management.

- Cradle-to-gate: it is an assessment of a partial product life cycle from raw material extraction or production (cradle) to the factory gate. There are two gates – before the product enters the factory and after it leaves the factory – which is defined as "cradle to gate 1" and "cradle to gate 2", respectively. In the former case, the manufacturing stage is not considered, while in the latter case, it is considered. In both cases, the use stage and disposal of the product are omitted. Cradle-to-gate assessments are sometimes the basis for environmental product declarations termed business-to-business. One example commonly used of the cradle-to-gate approach is the "cradle-to-site" for construction materials that covers the environmental impact from raw materials extractions and production and transportation to the construction site (RICS 2012).
- Gate-to-gate: it is the simplest option and includes the analysis only from reception of the raw materials to the end of production or the stage of manufacturing.

Step 4: Working out a Process Map

Process map is used for defining the system boundary. Companies shall include a process map in their inventory report (WBCSD/WRI 2011). A process map illustrates the services, materials and energy used in its life cycle. In includes:

- The defined life cycle stages as discussed above
- The generalized attributional processes in each stage
- The identified service, material and energy flows needed for each attributable process

- Any exclusions

The system boundary of the product life cycle shall exclude the GHG emissions associated with:

– Human energy inputs to processes and/or pre-processing (e.g. if fruit is picked by hand rather than by machinery)
– Transport of consumers to and from the point of retail purchase
– Transport of employees to and from their normal place of work
– Animals providing transport services

Taking an example adopted from Guide to PAS 2050 published in 2008 (Crown and Carbon Trust 2008), as shown in Fig. 2.10:

- Firstly, the scope is cradle-to-grave, where five life cycle stages have been defined for this croissant example, and they are raw materials, manufacture, distribution/retail, consumer use and disposal (assuming recycling not covered).

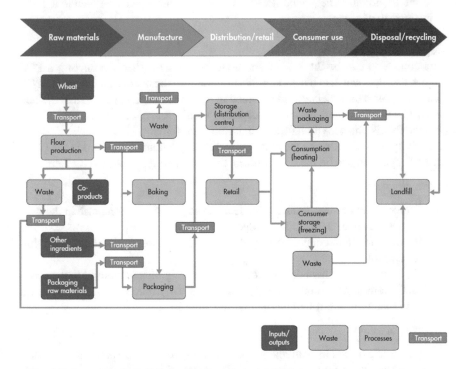

In this simplified example, a reliable and representative emission factor for wheat is assumed to exist, and therefore wheat production is not decomposed into its upstream activities (e.g. fertiliser production, transport and use; impact of land use change). Similarly, other ingredients and packaging are assumed to have reliable and representative emission data available. Although butter would be an important contributor to the product's overall footprint, for simplicity it is not included in detail in the calculations.

Fig. 2.10 Example of process map of croissants. (Adopted from *Guide to PAS 2050:2008* (Crown and Carbon Trust 2008))

- Secondly, under each life cycle stage, processes that contribute to GHG emission and removal are defined, in light blue box. For instance, under raw material stage, there is flour production, and under manufacture stage, there are baking and packaging. Transport within each stage or between different stages is also defined. Wheat as the raw material for flour needs to be delivered to the flour production site, while flour produced shall be delivered to the manufacturing site for baking.
- Thirdly, materials flows and materials input and output for each process are defined, and the amount of which is normally determined by using mass balances, especially if there are different raw materials, products, by-products and wastes. Energy consumptions including the electricity and fuel consumption from each process and those by transportation are also defined in this step.
- Lastly, there is a list of exclusions, as shown at the bottom of the process map. For instance, in this simple example, the carbon footprint of wheat is assumed to be known, and the GHG contribution of butter, which is an important ingredient to the croissant, is neglected. Reasons of the exclusion shall be elaborated and justified in this step.

Step 5: Collecting Data

Data collection can be the most resource-intensive step when conducting a PCF project, and data can also have a significant impact on the overall quality of the PCF measured. The data collected and recorded in relation to a PCF project shall include all GHG emissions and removals occurring within the system boundary of that product, that is, the activity data and emission factors.

Activity data includes all the material and energy amounts involved in the product's life cycle. Based on the data sources, there are two types of activity data:

- Primary data

  - Refer to direct measurements made internally or by someone else in the supply chain about the specific product's life cycle
  - Companies shall collect primary data for all processes under their ownership or control
  - For retailers or others that do not contribute a significant amount to the product's emissions, primary activity data is normally required for the processes and materials controlled by the first-tier supplier.

- Secondary data

  - Refer to external measurements that are not specific to the product, but rather represent an average or general measurement of similar processes or materials, e.g. industry reports or aggregated data from a trade association
  - Secondary data shall be used for inputs where primary activity data have not been obtained
  - According to PAS 2050:2011 (BSI 2011), preference shall be given to the use of cradle-to-gate information from a supplier over other secondary data.

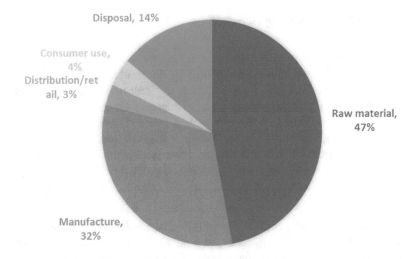

Fig. 2.11 Percentage of emission of different life cycle for croissant

During the data collection process, companies shall assess the data quality of activity data, emission factors and/or direct emissions data by using the data quality indicators. For significant processes, companies shall report a descriptive statement on the data sources, the data quality and any efforts taken to improve data quality.

Step 6: Calculating Results
Calculated results shall be reported in total emission in CO_2e per unit of analysis of the product under study, as well as the percentage of the emission by each life cycle stage defined in the system boundary. For the example above, the PCF of the croissant is 1.2 tonnes CO_2e per tonne of croissant, and contribution of emission from the different life cycle stage in percentage is shown in Fig. 2.11.

Step 7: Reporting
Reporting is crucial to ensure accountability and effective engagement with internal and external stakeholders. GHG Product Protocol (WBCSD/WRI 2011) has specific requirement on reporting, where readers could take reference from. Interpretation and evaluation of the results should also highlight opportunities to reduce or mitigate the climate change impact of the product or service under PCF study.

It also should be noted that in this book, biogenic emission is not covered and will not be discussed. For PCF calculation and reporting, biogenic and non-biogenic emissions are separated when applicable, as well as land-use change impacts. Interested party could also refer to the GHG Protocol for details.

Step 8: Assurance
Assurance is to make sure that the PCF calculation results and report are complete, accurate, consistent, transparent, relevant and without material misstatements. GHG Product Protocol required an independent assurance process, where assurer could not be the ones involved in the PCF project or the process in determining the PCF.

2.8.4   Carbon Labelling Schemes

Like company carbon reporting and disclosure as discussed in Sect. 2.7.2, product carbon labelling schemes serve the platform for companies to report the PCF of their products and to share with the stakeholders the information such as their effort in reducing the climate change impacts and to engage their stakeholders to reduce the PCF collectively. Based on the information released, there are normally two types of product carbon labelling schemes: one is the PCF itself and the other is the carbon reduction from PCF. In this section, a few labelling schemes will be introduced. It should be noted that carbon labelling on product carbon neutrality is not covered here, but in Chap. 5.

The UK
Carbon Trust in the UK launched the world's first product carbon footprint, the Carbon Trust Carbon Reduction Label, as a pilot in 2006, allowing companies to make independently verified claims related to the climate change impact of their products. It allows business sector to communicate with their consumers regarding:

- Their commitment to reduce their products' carbon footprint
- The PCF
- How consumers could contribute to reduce their own carbon footprint by preparing, using or disposing the product in the most efficient way

The Carbon Trust Carbon Reduction Label is not restricted to specific products or services. Carbon emission under PAS 2050 may be measured "business-to-consumer" from raw material to consumer use and finally disposal and/or recycling. It can also be measured as "business-to-business", covering only cradle-to-gate, that is, to a manufacturer.

Canada
Following the Carbon Reduction Label in the UK, other labelling schemes started to be launched by different organizations in Europe and North America. Examples include the CarbonCounted and Climatop, which were adopted in 2007 in Canada and in 2008 in Switzerland, respectively.

CarbonCounted is a Canadian not-for-profit organization. It allows businesses to calculate their GHG emissions using a third-party verified web-based GHG inventory system – Carbon Connect – and to help determine, manage and report the GHG inventories and product carbon footprints.

Japan
In Japan, the Minister of Economy, Trade and Industry (MEITI) initiated a 3-year national pilot project called Carbon Footprint of Products (CFP) in 2009 and officially launched "CFP Communication Programme", which was taken over by the Japan Environmental Management Association for Industry (JEMAI) in 2012.

The pilot project of carbon labelling scheme was voluntary for retailers to attach the Carbon Footprint Label to their products to encourage organizations to reduce their GHG emissions and to allow customers to choose daily products to achieve low carbon lifestyle. Calculation of PCF follows the LCA principles from raw material acquisition to disposal of the commodities. The label as shown in Fig. 2.12 includes an image of a lead weight with the letters CO_2 in the centre, with the attached carbon weight of the product in bold letters above. The attached carbon weight value is an approximation of the amount of carbon released across the entire life cycle of the product.

South Korea
Following Japan, Korea Environmental Industry and Technology Institute (KEITI) introduced a voluntary programme to certify carbon content in consumer goods in 2009. It consists of a two-step certificate: "Carbon Emissions Certificate" which states the PCF, illustrated by a CO_2 image, and "Low-Carbon Product Certificate", which verifies that low levels of carbon have been emitted in the production of the product, with the product's carbon footprint displayed, as shown in Fig. 2.13.

There are two criteria for Low-Carbon Product Certificate:

- Criteria 1 on average carbon emissions: Meet the average value of carbon emissions among previous carbon emissions-certified products in the same product category.
- Criteria 2 on carbon reduction rate: Meet 4.24% carbon reduction rate per 3 years which is generated from "target scheme for energy and greenhouse gas emission".

It covers all products and services, excluding agricultural products, fishery products, livestock products, medical products, and exported products.

Table 2.10 summarizes the different PCF labelling schemes.

Fig. 2.12 Japan's CFP mark (logo) (https://www.cfp-japan.jp)

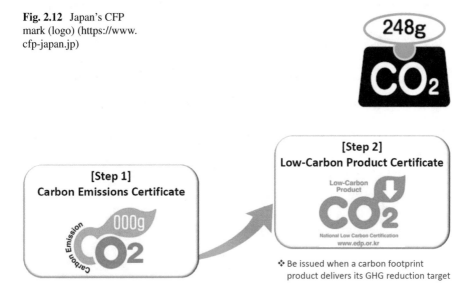

Fig. 2.13 South Korea's carbon footprint labelling scheme. (KEITI 2013)

Table 2.10 Overview of the different product carbon footprint labelling schemes

Country	Scheme name	Year	Accounting method	Implementing agency
UK	Carbon Reduction Label	2006	PAS 2050	Carbon Trust
Canada	Carbon Connect/ CarbonCounted™ Standards	2007	Unspecified LCA	CarbonCounted
USA	Certified CarbonFree	2007	PAS 2050 ISO 14044	Carbonfund.org
	Climate Conscious Carbon Label	2007	unspecified LCA	The Climate Conservancy
	Pilot Californian carbon label	2009	unspecified LCA	California State Senate Carbon Labeling Act 2008
Switzerland	Climatop	2008	GHG protocol ISO 14064	Climatop
Japan	Carbon Footprint of Products	2009	ISO 14040,14044 ISO 14067	Minister of Economy, Trade and Industry (METI)and Japan Environmental Management Association for Industry (JEMAI) since 2012
South Korea	Carbon Footprint Certification Label /Low Carbon Product Certificate	2009	PAS 2050	The Korea Environmental Industry and Technology Institute (KEITI)
Thailand	Thailand Carbon Reduction Label	2009	PAS 2050	Thailand Greenhouse Gas Management Organisation
Taiwan	Taiwan BSI Product Carbon Footprint	2010	PAS 2050	British Standards institution (BSI)
Australia	Carbon Reduction Label (from Carbon Trust)	2010	PAS 2050	Planet Ark
France	I' Indice Carbone Label	2011	Methode Bilan Carbone®	Casino France
China	Product Carbon Footprint Label	2013	Cradle-to-grave LCA	China Quality Mark Certification Group

References

BSI (2011) PAS 2050:2011 Specification for the assessment of the life cycle greenhouse gas emissions of goods and services, 2011

California Climate Action Registry (2009) General Reporting Protocol Reporting Entity-wide Greenhouse Gas Emissions

Carbon Trust (2006) Carbon footprints in the supply chain: the next step for business

Carbon Trust (2008) Product carbon footprinting: the new business opportunity Experience from leading companies

Crown and Carbon Trust (2008) Guide to PAS 2050 How to assess the carbon footprint of goods and services, 2008

DEFRA (2018) UK Government emission conversion factors for greenhouse gas company reporting

EMSD/EPD (2010) Guidelines to Account for and Report on Greenhouse Gas Emissions and Removals for Buildings (Commercial, residential or Institutional Purposes) in Hong Kong

IPCC (2006) 2006 IPCC Guidelines for National Greenhouse Gas Inventories, Prepared by the National. In: Eggleston HS, Buendia L, Miwa K, Ngara T, Tanabe K (eds) Greenhouse Gas Inventories Programme. IGES, Japan

ISO (2018a) ISO 14064-1: Greenhouse Gases- Part 1: Specification with guidance at the organization level for quantification and reporting of greenhouse gas emissions and removals

ISO (2018b) ISO 14067:2018: Greenhouse gases — Carbon footprint of products — Requirements and guidelines for quantification

KEITI (2013) Current Status and Updated Issues of Carbon Footprint in Korea, International Workshop on Future Utilization of Visualized Information of Environmental Impacts in Product Life Cycle & Corporate Value Chain at Tokyo, Japan, 27th Feb 2013, Carbon Management Office of KEITI

RICS (2012) Methodology to calculate embodied carbon of materials. RICS Information Paper. IP32/2012

Steer A, Hanson C (2015) Biofuels are not a green alternative to fossil fuels. From theguardian.com 29 Jan 2015

The Climate Registry (2008) General Reporting Protocol version 1.1: Accurate, transparent, and consistent measurement of greenhouse gases across North America

WBCSD/WRI (2010) The GHG Protocol Corporate Accounting and Reporting Standard

WBCSD/WRI (2011) The GHG Protocol Product Life Cycle Accounting and Reporting Standard

Chapter 3
Carbon Trading and Offsetting

3.1 Emergence of Carbon as an Asset

According to Financial Accounting Standards Board's (FASB) Concept Statement 6, an asset has three essential characteristics:

(a) It embodies a probable future benefits that involves a capacity to contribute directly or indirectly to future net cash inflows.
(b) The company can obtain the benefit and control others' access to it.
(c) The transaction or other event giving rise to the company's right to or control the benefit has already occurred.

Carbon has been emerged as an asset, which could be illustrated from two aspects.

Firstly, as already discussed in Sect. 1.5.4, Paris Agreement endorsed a long-term goal to limit global temperature rise to well below 2 °C and to pursue efforts towards 1.5 °C. In order to limit warming to 1.5 °C, net global CO_2 emissions need to fall by about 45% from 2010 levels by 2030 and reach "net zero" by around 2050. For 2 °C, CO_2 emissions will need to decline by about 20% by 2030 and reach net zero around 2075 (Fig. 3.1). All pathways that limit global warming to 1.5 °C project the use of carbon dioxide removal (CDR) on the order of 100–1,000 $GtCO_2$ over the twenty-first century (IPCC 2018).

Secondly, the global carbon abatement burden, however, should be shared by both developed and developing countries. As shown in Fig. 3.2, rich countries, who, on a per capita basis, enjoy a disproportionate fraction of the atmospheric commons, need to reduce their emissions while at the same time allowing developing countries to catch up by increasing their per capita emissions. This "convergence" implies greater equity in emissions across nations while at the same time emissions targets for the world as a whole are met. This paves the ground for the differentiated responsibilities when pursuing collective global goal of carbon reduction. It also worth

© Springer Nature Switzerland AG 2020
S. W. W. Zhou, *Carbon Management for a Sustainable Environment*,
https://doi.org/10.1007/978-3-030-35062-8_3

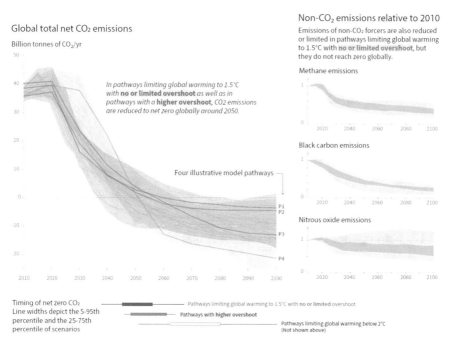

Fig. 3.1 Global emission pathway characteristics (IPCC 2018)

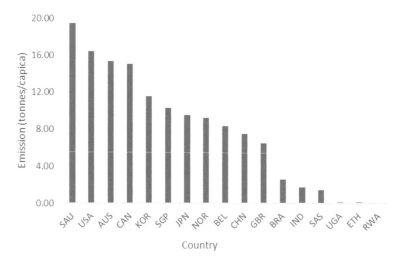

Fig. 3.2 Emission per capita for selected countries (2014 data from data.worldbank.org)

noting that China, although still a developing country, has already caught up with some developed countries in both economic development and emission reduction progress.

Carbon has been emerged as an intangible asset, which is characterized as a non-monetary asset that cannot be seen, touched or physically measured and which is created through time and/or effort. Carbon assets include:

1. Offsets (carbon credits)
2. Facilities that generate performance credits
3. Allowances

3.2 Emission Trading Benefits

Trade involves the transfer of the ownership of goods or services from one person or entity to another in exchange for other goods or services or for money. A network that allows trade is called a market. Emissions trading, or cap-and-trade, is a market-based approach to controlling pollution by providing economic incentives for achieving reductions in the emissions of pollutants. In contrast to command-and-control environmental regulations such as best available technology standards and government subsidies, cap-and-trade programs are a type of flexible environmental regulation that allows organizations to decide how best to meet policy targets.

Emission trading has been developed and successfully implemented for around 30 years. The earliest and most successful example of implementing cap-and-trade is the USA. In 1990, the USA revised the Clean Air Act to implement emissions trading mechanisms for sulphur dioxide (SO_2) and nitride (NOX) to address the problem of acid rain. To meet the emission standard, the company is allowed to purchase the emission allowance from the market, the price of which is only one-tenth of the original fine of US$2000 per tonne as penalty of excessive emission in the command-and-control system.

Benefits of carbon trading include:

- Effectiveness
  Cap-and-trade has proven its effectiveness in the USA through the acid rain program, where it quickly and effectively reduced pollution levels at a far lower cost than expected. The EU Emissions Trading System has shown that cap-and-trade can be extended to carbon and in doing so creates a price on carbon that drives emission reductions. Constrained by a cap, emitters will innovate. Emissions trading incentivizes innovation and identifies lowest-cost solutions to make businesses more sustainable.
- Efficiency
  Emission trading could reduce the reduction cost (Chan et al. 2017). By allowing emission sources with high abatement costs to offset higher on-site emissions by purchasing additional reductions from other, lower-cost polluters, they assert trade in pollution allowances reduces the total cost of achieving a given reduction in aggregate emissions.
- Integration

Emissions trading can overcome the regional disparities, be able to better respond to economic fluctuations than other policy tools and provide a global response to a global challenge. Finance and technology can be flown into the emerging market via the trading scheme.

Carbon Tax

Besides cap-and-trade, the other approach to price carbon is carbon tax. A carbon tax directly establishes a tax levy on carbon emissions, which is factored into the price of goods and services based on their carbon content. Both approaches can reduce emissions by encouraging the lowest-cost emission reductions and encourage investors and entrepreneurs to develop new low-carbon technologies.

A carbon tax offers stable carbon prices, so carbon is not a potential risk with fluctuating regulatory costs when making any business decision. In addition, if emission reductions are cheaper than expected – for example, an economic downturn causes emissions to fall – then a tax provides a continuing price signal, whereas cap-and-trade does not encourage reductions beyond the emissions target. Nevertheless, cap-and-trade offers more effective and efficient emission reduction, with the good design program such as the following: more stringent emission caps; price floors and ceilings can avoid volatility; emissions allowances can be auctioned instead of given away; etc.

3.3   Current Carbon Market

3.3.1   Types of Carbon Market

There are two types of carbon market – compliance market and voluntary market. The compliance market is used by companies and governments that by law have to account for their GHG emissions. It is regulated by mandatory national, regional or international carbon reduction regimes. On the voluntary market the trade of carbon credits is on a voluntarily basis.

Under compliance market, there are two approaches to emissions trading: cap-and-trade and baseline-and-credit. In cap-and-trade, the regulatory authority sets an aggregate cap on emissions and then divides the cap into a number of tradable permits (frequently called allowances). Firms then trade the allowances, establishing a market price. In baseline-and-credit, there is no explicit cap on aggregate emissions. Instead, each firm has the right to emit a certain baseline level of emissions. This baseline may be derived from historical emissions or from a performance standard that specifies the permitted ratio of emissions to output. Firms create emission reduction credits when its emission is below the baseline. These credits may be banked or sold to firms who exceed their baselines. Table 3.1 summarizes the different types of carbon market and its approaches.

Table 3.1 Types of carbon market

Carbon market	Carbon credit	Trading scheme
Allowance market (cap-and-trade)	EUA	EU Emissions Trading System
	AAU	International Emissions Trading
Project-based (baseline-and-credit)	CER	Clean Development Mechanism
	ERU	Joint Implementation
Voluntary market	VER	Voluntary Carbon Offsets
	VCU	Voluntary Carbon Unit (VCS Standard)

3.3.2   World Carbon Market

According to the World Bank's report (World Bank and Ecofys 2018), by 2018, 51 carbon pricing initiatives have been implemented or are scheduled for implementation, as shown in Fig. 3.3. This consists of 25 emissions trading systems (ETSs), mostly located in subnational jurisdictions, and 26 carbon taxes primarily implemented on a national level. These carbon pricing initiatives would cover 11 gigatonnes of carbon dioxide equivalent (GtCO₂e) or about 20 percent of global greenhouse gas (GHG) emissions. In 2018, the total value of ETSs and carbon taxes is US$82 billion, representing a 56 percent increase compared to the 2017 value of US$52 billion.

Table 3.2 compares the different carbon trading schemes from the first established EU Emission Trading System (ETS) in 2005 to the newly launched China's national emission trading scheme in 2017.

One has to be noted that cap-and-trade is a controversial topic with a lot of scandals. Students and audience are encouraged to think deeply on the aspects such as how to define and set the cap, how the system could benefit the poor and how to achieve real reduction. To stimulate your thinking, it is recommended to watch the video called Story of Cap and Trade produced by The Story of Stuff Project.

3.3.3   Project-Based Mechanisms

Under project-based mechanism, it uses baseline-and-credit scheme for carbon trading. A baseline, which can be business as usual or some proportion thereof, is set, and credits are created for activities whose emission below the baseline and activities whose emission above the baseline have to buy such credits. The ability to generate credits from emission reductions relative to baseline and the pressure to avoid having to buy permits for emissions in excess of the baseline provide incentives for participants to find lower emission production processes.

Joint Implementation (JI) and Clean Development Mechanisms (CDM) are the two project-based mechanisms which feed the carbon market. JI enables industrialized countries to carry out joint implementation projects with other developed countries. CDM involves investment in sustainable development projects that reduce

Table 3.2 World carbon trading schemes comparison

Market	Development	Reduction targets	Sector covered	Tools and mechanisms
EU ETS	EU ETS directive adopted in 2003; linking directive with Kyoto Protocol in 2004; EU EU ETS Phase I (2005–2007) launched in 2005; Phase II (2008–2012) Kyoto commitment period; adoption of the 2020 EU energy and climate package with a revised directive for the Phase III (2013–2020) in 2009	20% below 1990 levels (or 13% below 2005 levels) by 2020; long-term objective is to reduce the EU's emissions to 80–95% below 1990 levels by 2050	Power and heat generation; industrial processes; production of cement, glass, lime, bricks, ceramics, pulp, paper and board; commercial aviation; CCS networks; production of petrochemicals, ammonia, non-ferrous metals, gypsum, aluminium and nitric, adipic and glyoxylic acid	Mandatory for all EU members, plus Iceland, Liechtenstein and Norway, covering about half of total EU carbon emissions Free allowance allocation, offsets, banking, market stability reserve (2019)
New Zealand ETS	NZ ETS launched in 2008. Originally designed to provide access to international credits. However, the NZ ETS restricted the use of CERs, ERUs, and RMUs as of 1 June 2015, making it a domestic-only system	5% below 1990's levels by 2020; 30% below 2005's level (11% below 1990's level) by 2030; 50% below 1990's levels by 2050	Unit surrender and emissions reporting: forestry, stationary energy, transport, industrial processes, synthetic GHGs and waste. Emissions reporting only: biological emissions from agriculture	Intensity-based allocation for the industrial sector (26 eligible activities): 90% free allocation for highly emissions-intensive and trade exposed activities
Tokyo cap-and-trade	The Tokyo Metropolitan Government Cap-and-Trade Program (TMG ETS) is Japan's first mandatory ETS launched in 2010	25% reduction from 2000 levels by 2020; 30% by 2030	Commercial and industrial sectors; emission limits set for large factories and offices	The absolute cap is set at the facility level that aggregates to a Tokyo-wide cap
RGGI	The Regional Greenhouse Gas Initiative (RGGI) was the first mandatory trading program that caps emissions in the USA through state-coordinated cap-and-trade programs, after the failure of federal cap-and-trade legislation in 2010	A regional target of more than 50% reduction in CO_2 emissions from the power sector by 2020 relative to 2005 levels	Fossil fuel electric generating units in nine Northeastern and Mid-Atlantic states	Majority of CO_2 allowances are issued by each state and distributed through quarterly, regional CO_2 allowance auctions using a "single-round, sealed-bid uniform-price" format
China ETS	Approved seven pilot tests of carbon trading in 2011 and launched national emissions trading scheme in Dec 2017	Linked to NDC carbon intensity target, no independent ETS target	1,700 coal- and natural gas-based power companies	Integration of cap, coverage and allocation (Qi and Cheng 2018)

Fig. 3.3   The world carbon market 2018 (World Bank and Ecofys 2018)

emissions in developing countries. Under CDM, emission-reduction projects in developing countries can earn certified emission reduction credits (CERs). These saleable credits can be used by industrialized countries to meet a part of their emission reduction targets under the Kyoto Protocol.

Figure 3.4 illustrates overall mechanism of CDM projects. CDM projects should be hosted in the non-Annex I countries or developing countries (like China, India, etc.) which adhere to UNFCCC conventions and establish a National Designated Authority. CDM projects are co-developed by both a host country and an Annex-I country.

Companies in Annex I countries could purchase the carbon credits generated from the CDM projects developed in host countries. It offers the companies in Annex I countries flexibility in meeting their reduction targets in a reduced cost and also provide an investment opportunity, while host countries could get additional financing for these projects, generate more jobs or infrastructure development opportunities and at the same time mitigate local emissions with more advanced technologies and solutions.

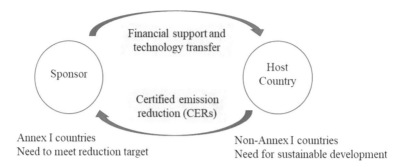

Fig. 3.4 Overview of Clean Development Mechanism (CDM)

3.3.4   Voluntary Market

The voluntary market generally applies to companies, individuals and other entities and activities not subject to mandatory limitations but wish to offset GHG emissions. Voluntary market is very small compared to the regulatory market but has been growing quickly. According to Ecosystem Marketplace's report (Ecosystem Marketplace 2018), since the voluntary carbon markets picked up in the late 2000's, offset issuances (supply of offset) and retirements (demand of offset) have both increased dramatically. In 2018, offset issuances (62.9 MtCO$_2$e) and retirements (42.8 MtCO$_2$e) reached record-highs.

Buyers include companies that buy offsets for their own operations, companies that buy offsets on behalf of their customers (e.g. airlines and travel agents, automobile and petroleum companies), events (e.g. 2010 World Cup) and individuals. Sellers include retailers and wholesalers who buy and resell offsets and project developers who develop GHG abating activities and sometimes sell direct.

Why companies want to purchase voluntary carbon offsets:

- Climate justice – addressing inequity concerns
- Incentivize GHG reduction with broad participation
- Cost-effectiveness – more reduction with less funds
- Innovation and experimentation
- Preparation for future regulated market
- Corporate goodwill

The carbon credits generated from voluntary markets can only be used in the voluntary, while credits generated from compliance market (e.g. CERs) can be traded in both compliance and voluntary markets. The details on how credits are generated and which standard to adopt will be discussed in the following two sections.

3.4   Carbon Offsetting

3.4.1   What Is Carbon Offset

A carbon offset is a reduction in emissions of carbon or GHGs made in order to compensate for or to offset an emission made elsewhere. In other words, offset projects reduce GHG emissions outside of an entity's organizational boundary, and that reduction is more cost-effective than what it would take to reduce the same amount of emissions within the entity's boundaries. Carbon offsetting is one of the means to achieve carbon reduction. But one should be noted that by offsetting there is no real reduction within the entity's boundary.

Carbon offsets are measured in metric tonnes of carbon dioxide-equivalent (CO_2e) and may represent six GHGs. Offsets are typically achieved through financial support of projects that reduce the emission of greenhouse gases in the short- or long-term. Offsets that are produced under a voluntary standard are called Verified Emission Reductions (VERs). Offsets that are generated under the Clean Development Mechanism are called Certified Emission Reductions (CERs). Buyers in the voluntary market can choose to buy CERs or VERs.

3.4.2   Types of Carbon Offset Projects

There are six broad types of carbon offsetting projects, namely:

- Biological sequestration
- Industrial gases
- Methane capture and utilization
- Energy efficiency
- Renewable energy
- Carbon capture and storage (adopted by UNFCCC at CMP7 in 2011)

Biological Sequestration
Land biological sequestration refers to the absorption of carbon dioxide from the atmosphere through photosynthetic processes by plants or trees and subsequent fixation into soils, plants or trees. Under the United Nations Framework Convention on Climate Change (UNFCCC), any process, activity or mechanism which removes a greenhouse gas from the atmosphere is referred to as a carbon sink. Human activities impact terrestrial sinks, through land use, land-use change and forestry (LULUCF) activities. Mitigation can be achieved through LULUCF activities that increase the removals of GHGs from the atmosphere or decrease emissions by sources leading to an accumulation of carbon sinks.

There are three types of LULUCF projects (Kollmuss et al. 2008)

- Those that avoid emissions via conservation of existing carbon sinks (*avoided deforestation*), called "reducing emissions from deforestation and degradation" (REDD)
- Those that increase carbon storage by sequestration:

  - *Afforestation* – is the process of creating forests on land that was previously unforested, typically for longer than a generation
  - *Reforestation* – is the process of restoring forests on land that was once forested

- Those that increase carbon storage by *soil management* techniques

Project: Afforestation and Reforestation on Degraded Lands in Northwest Sichuan

It is the world's first afforestation and reforestation project that is registered under the Clean Development Mechanism (CDM-AR) and simultaneously developed according to the most stringent international standards for "climate, community and biodiversity" (CCB). By integrating CCB standards with CDM protocol, this project pioneers both in China's forest management and in delivering measurable sustainability benefits to the local community and biodiversity.

The project, strongly supported by China's Ministry of Forestry and international and local environmental non-governmental organizations (NGOs), was successfully validated by and registered with the CDM-Executive Board under the United Nations on 16 November 2009 after 5 years of comprehensive research and preparation work. Totally over 2,250 hectares, this forest restoration project spans five impoverished counties in a district badly stricken by the Sichuan Earthquake in May 2008.

The project sites are degraded lands in the north-western highlands of Sichuan Province in China. As the sites are located in the buffer zones of important nature conservation areas, the project helps preserve the habitat of many rare species in one of the world's biodiversity hotspots.

More info refer to Project Design Document from: https://cdm.unfccc.int

Compared with other offsetting projects, tree planting projects are more acceptable by the buyers; however, there are many controversial issues (David Suzuki Foundation 2009), and a few are listed here:

- Methodological challenge – without an accurate understanding of the carbon inventory, it is difficult to quantify what the net carbon benefit of planting a tree is.

- Timing – as trees will be mature in many years, the credits selling on the assumption that the project will materialise and succeed in future, makes it a "forward selling" project.
- Permanence – trees provide temporary carbon storage as part of the normal cycle of carbon exchange between forests and the atmosphere. Trees can easily release carbon into the atmosphere through fire, disease, climatic changes, natural decay and timber harvesting. A well-told story was how Coldplay's carbon offsetting tree planting projects died in India, due to the lack of further funding and proper management (Dhillon and Harnden 2006).
- Biodiversity – in order to cut costs, some tree-planting projects introduce fast-growing invasive species that end up damaging native forests and reducing biodiversity.

It is noted that tree planting projects need not only good design and planning; like all other types of offsetting projects, tree planting also needs intensive monitoring and management, and it is recommended to follow a more stringent standard, such as CCB standard as mentioned in the case above, to better manage the project risk and contribute more to the community.

Industrial Gases Projects

Some industrial gases have very high GWP, for instance:

- Nitrous oxide (N_2O) (GWP = 310), released in production of adipic and nitric acids
- HFC-134a and HFC-23 (GWP 1300 and 11,700, respectively), which are by-products from production of refrigerants (HCFC-22)
- PFCs (mainly CF_4 and C_2F_6; GWP 6500 and 9200, respectively) released in the aluminium industry
- SF6, the most powerful greenhouse gas used in power substations and LCD screen production (GWP = 23,900)

Because of their high GWP, to destruct these gases is an effective way to reduce carbon emission. Industrial gases project is the cheapest and most efficient type of projects in generating carbon offset; however, they do not contribute to the path to a low-carbon economy and deliver few additional environmental and social benefits. Some reports have indicated that the creation of an offset market for HFC-23 gases has created perverse incentives in China and India to start building new HCFC-22 facilities to increase revenue from offsets, as HFC-23 is created as a by-product during HCFC-22 production. Although CDM hasn't banned the establishment of new HCFC-22 plants, European Union Emissions Trading System banned the use of the use of HFC-23 certified emission reduction (CER) credits since 1 May 2013. Other carbon markets have followed suit, resulting in the collapse of the HFC-23 credit market.

Methane Capture and Utilization

Methane is produced and emitted by landfills, during wastewater treatment, in natural gas and petroleum systems, by agriculture (livestock and rice cultivation) and during coal mining. There are two types of methane projects. The first type captures and fares methane. Through combustion, methane gas is turned into less potent CO_2 and H_2O. Examples of such projects include the capture and faring of landfill gas and of coal mining gas. The second type of project captures methane and uses it to produce either hot water or electricity. Such projects include those that capture and purify methane in wastewater treatment plants or landfills and use it for electricity production or the production of another form of energy.

Biofuel plants that use agricultural or forestry waste to produce electricity also utilize methane. Organic matter is anaerobically digested, and the resulting methane is used to produce electricity or heat. But such biofuel projects are considered renewable energy projects rather than methane capture projects. An example of food waste biofuel facility will be discussed in detail in Chap. 7.

Energy Efficiency Projects

Energy efficiency (EE) can be improved through changes in three different categories (Hinostroza et al. 2007):

- Process and design change

  - A complete or partial change to the elemental processes, e.g. changing the recipe of a cement blend so it requires less heat per output unit or changing the orientation and natural ventilation of a building

- Technological change

  - This includes equipment upgrade and installation of new hardware based on a more efficient technology, e.g. better insulation for buildings, more efficient household appliances, replacing old boilers, changing burners, etc.

- Fuel switching, distributed generation

  - Not typically considered EE measures, but supply-side projects, which reduce requirement for fossil fuels and improve overall efficiency

Table 3.3 gives some examples of EE projects. It should be noted that although end-user EE project is one of the most promising type of projects for reducing emission and increasing energy security in developing countries, EE projects have different owners which may be geographically dispersed, which make this type of project difficult to apply and manage, and may also have a high cost in administrative work (Figueres and Philips 2007).

Renewable Energy

Renewable energy (RE) projects include bioenergy, geothermal, hydropower, ocean energy, solar photovoltaics (PV), solar thermal and wind power. By the end of 2016, renewable energy accounted for an estimated 18.2% of global total final energy consumption (REN21 2018), and by January 2018, over 100 cities got at least 70% of their electricity from renewable sources (Carbon Disclosure Project 2018).

Renewable energy projects are crucial in term-term low carbon energy transformation by providing energy for heating and cooling, power generation and transportation with no or only marginal direct and indirect CO_2 and other GHG emissions.

Table 3.3 Examples of EE projects

Sector	Examples of project
Households and commercial buildings	New development project with various energy-saving technologies; this can be a stand-alone project if the size is big enough. New development of low-carbon community with small building owners. Retrofitting projects for existing buildings. Building energy efficiency is well covered under Green Building Labelling schemes in different countries
Industry	Combined heating process, steam reduction, improved boiler design, modification of clinker cooler, pumps vans and compressors optimization, etc.
Transportation	EE can be achieved mostly through centralised action, such as fuel switching by car owners, vehicles replacement and encouraging public transportation by individuals, private or public entities, etc.
Agriculture	Agriculture similar to industrial, but a smaller scale; improving agriculture vehicles and tractors' fuel efficiency
Supply	Conversion of thermal power stations from open cycle to combined cycle, thermal efficiency improvements, waste heat recovery and co-generation (e.g. in cement factories, sugar factories, steel plants)

Renewable energy could improve supply security. Figure 3.5 shows that renewable energy sources are inexhaustible and their natural availability is 3,000 times higher than current global annual energy consumption. Even the current technical potential for renewables use is six times higher. In addition, renewable energy sources are available in most regions of the world, not just in specific geographical areas as is the case with fossil and nuclear resources, which means the price is less dependent and fluctuated compared with fossil fuels (Wolfgang et al. 2007).

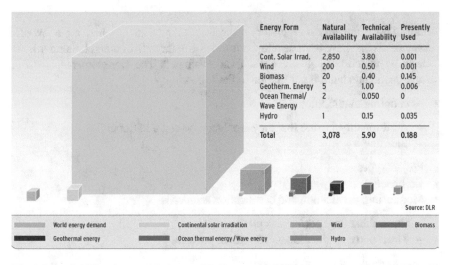

Fig. 3.5 Availability of renewable energy (Wolfgang et al. 2007). *Rear cubes*: The natural availability of renewable energy is extraordinarily large. *Front cubes*: The technically available energy in the form of electrcity, heat and cheical energy carriers exceeds the present-day energy demand (grey cube, left) by a factor of six

However, as one type of carbon offsetting projects, RE projects have a few barriers, including the higher capital costs and lower retune of the certified carbon credit, as RE projects typically reduce emissions of CO_2, which has a global warming potential of 1, and many RE projects are relatively small scale.

Carbon Capture and Storage

Carbon dioxide capture and storage (CCS) means the capture and transport of carbon dioxide from anthropogenic sources of emissions such as power plants, steel, cement, and fertilizer factories and other industrial facilities and the injection of the captured carbon dioxide into an underground geological storage site for long-term isolation from the atmosphere. The process is currently being demonstrated worldwide at commercial scales, but without more experience, the role of CCS in meeting future climate goals will remain uncertain – especially in emerging economies (Almendra et al. 2011).

CCS, specifically carbon dioxide capture and storage in geological formations was adopted as Clean Development Mechanism (CDM) project activities under the Kyoto Protocol at the December 2011 decision by the meetings of the United Nations Framework Convention on Climate Change (UNFCCC) in Durban, South Africa, as was the culmination of many years of international negotiation.

It should be noted that CCS project is not applied to the sectoral scopes under Verified Carbon Standard (VCS), which is a voluntary carbon offset standard we are going to discuss in the next section.

3.4.3   Carbon Offset Standards

While CDM is the largest regulatory project-based mechanism, there are a few offset standards that have been developed for the voluntary offset market. Standards set criteria by which projects are chosen and evaluated. The key elements of offset standards include but are not limited to the following:

- Accounting methodologies

  - Additionality, ensure that they are "genuine and additional"
  - Baseline

- Project types
- Co-benefits
- Monitoring, verification and certification standards

  - Ensure that they perform as predicted
  - Third-party verification and approval process

- Registration and enforcement system

  - Ensure no double-selling, clarify ownership
  - Enable trading of offsets

Additionality

Additionality is a very important concept to carbon offsetting project. It means the project has additional value, and it would not happen if the project was not implemented as an offset project. Additionality test provides the tool to evaluate the project. For instance, CDM additionality tool is listed below:

- Regulatory surplus – if the project is implemented to fulfil official policies, regulations or industry standards, it cannot be considered additional. It must be beyond the regulatory requirement.
- Investment analysis – revenue from carbon offset could bring positive return of investment.
- Barrier analysis – if the project succeeds in overcoming significant non-financial barriers that the business-as-usual alternative would not have had to face, the project is considered additional.
- Common practice test – if the project employs technologies that are very commonly used, it might not be additional because it is likely that the carbon offset benefits do not play a decisive role in making the project viable.

Co-benefits

Co-benefits is defined as the social and environmental benefits or sustainability development benefits from the offsetting projects. Like the afforestation and reforestation in Northwest Sichuan we discussed above, this project could provide jobs to the local community, protect the panda's habitat and reduce water erosion and landslides. Co-benefits is important when an offsetting project is designed and planned. Stakeholders of the project should be publicly consulted to better understand project site's situation and to consolidate stakeholders' feedbacks, advices and suggestions.

Registry and Double-Counting

Carbon offset registries keep track of offsets and clarify ownership of offsets. A serial number is assigned to each verified offset. When an offset is sold, the serial number and carbon credit is transferred from the account of the seller to an account of the buyer. The registry then retires the serial number so that the credit cannot be resold. It avoids the double-counting, where same carbon credits are owned by different buyers and use for offset for more than once. CERs are registered at CDM registry. There is no one single registry for the voluntary market. Corporate should keep their own carbon credit inventory to retire the credits and avoid double-counting.

Table 3.4 compares different offset standards.

The GHG Protocol for Project Accounting is an offset accounting protocol for quantifying and reporting GHG emission reductions from GHG mitigation projects and does not focus on verification, enforcement or co-benefits. ISO 14064-2 provides general guidance and does not prescribe specific requirements. It suggests that additionality be taken into account but does not require a specific tool or additionality test to be used. Similarly, it does not focus on co-benefits.

Table 3.4 Summary of comparison among different carbon offset standards

Standards	Additionality	Project types	Co-benefits	Third-party verification/ separate approval process	Registry
CDM	Project specific, additionality test	All except nuclear energy, new HCFC-22 facilities and avoided deforestation	Depending on host countries Stakeholder consultation is required	Yes/yes	CDM registry, approved by CDM-Executive Board
GHG Protocol for Project Accounting	No requirement	Any	Not applicable	Not applicable	Not applicable
ISO 14064-2	No requirement	Any	Not applicable	Strongly recommends the use of third-party auditors, but it is a requirement to do so only if they are reported publicly	Not applicable
Gold Standard	Same as CDM	Renewable energy (including methane-to-energy projects) and end-use energy efficiency No large hydro above 15 MW	Must demonstrate social and environmental benefits, very strict consultation requirement	Yes/yes	Markit Environmental Registry, Water Benefit Standard Registry
VCS	Project-based test	15 Sectoral scope, does not include new HCFC-22 and CCS projects	No extra requirement except for legal environmental compliance; consultation is required	Yes/no	Multi-registry system

(continued)

Table 3.4 (continued)

Standards	Additionality	Project types	Co-benefits	Third-party verification/ separate approval process	Registry
VER+	Same as CDM	Any except any HFC projects, nuclear power projects and hydropower projects exceeding 80 MW	Local stakeholder consultation required only	Yes/no	Blue Registry of TÜV SÜD
CCBS	Project specific, individual methods	LULUCF	Must generate positive environmental, social and economic impacts. Stakeholder involvement is required and must be documented	Yes/no	Verra tracks all CCB projects

The Gold Standard (GS) is a voluntary carbon offset standard for renewable energy, energy efficiency, waste management and land-use and forests projects. It was launched in 2006 and has been endorsed by over 60 environmental and development NGOs. The GS can be applied to voluntary offset projects and to CDM projects. Founded under the leadership of World Wildlife Fund (WWF), it differentiates itself by requiring the sustainable development goals of offsetting projects, i.e. beyond offsetting emissions, also contributing to the economic and social welfare and development of the people of project site. These sustainability requirements are more stringent than those of VCS or CDM.

The Verified Carbon Standard (VCS) is a standard founded in part by the International Emissions Trading Association, a consortium whose members include major energy and chemical companies, banks and law firms. Unlike the GS or CDM, VCS has no requirement that its carbon offset project have additional social benefits, allowing for a wider range of projects. VCS is broadly supported by the carbon offset industry (project developers, large offset buyers, verifiers and projects consultants). Verified Carbon Units (VCUs) are issued under VCS. The VCS registry system is a multi-registry system, currently comprised of two registry service providers (APX and HIS Markit) and a central project database.

The VER+ closely follows the Kyoto Protocol's project-based mechanisms (CDM and JI). The VER+ standard was developed by TÜV SÜD, an auditor (Designated Operational Entity – DOE) for the validation and verification of CDM projects. It was designed for project developers who have projects that cannot be implemented under CDM yet who want to use very similar procedures as the CDM. The VER+ was launched in mid-2007.

The Climate, Community & Biodiversity Standards (CCB Standards) is a project design standard that offers rules and guidance for project design and development. It is intended to be applied early on during a project's design phase to ensure robust project design and local community and biodiversity benefits. It does not verify quantified carbon offsets, nor does it provide a registry. The CCB Standards focuses exclusively on land-based bio-sequestration and mitigation projects and requires social and environmental benefits from such projects.

For more detailed discussion of different carbon offsetting projects and their comparisons, audience can refer to WWF publication (Kollmuss et al. 2008).

3.5   Carbon Credits

In this section, we will discuss how carbon credits are produced and, as a sustainability consultant or corporate sustainability personnel, how one can develop an offsetting project or choose the right carbon credits to fulfill the company's carbon targets.

3.5.1   How Carbon Credits Are Born

As we discussed in the last section, different carbon offset standards may have different project development processes. Here we use CDM project as a framework reference to illustrate how carbon credits are issued and verified.

As shown in Fig. 3.6, at the project design stage, the project developer should prepare the Project Design Document (PDD), which includes the project concept, the baseline, monitoring methodologies and outcomes of stakeholder consultations. The

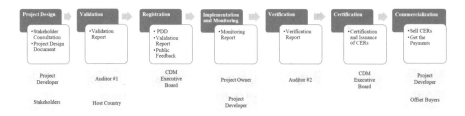

Fig. 3.6 The CDM Project Cycle and key players involved

project developer will then submit the PDD to the third-party auditor (i.e. auditor #1) for project validation. Under CDM, these auditors are called Designated Operational Entities or DOEs. Host country approval must be obtained for CDM projects. PDD, validation report and public feedbacks are submitted to the CDM Executive Board (CDM EB) for review and project registration.

Project owner – the operator and owner of the physical installation where the emission reduction project takes place can be any private person, company or other organization and can implement the project anytime in the project cycle, but it is recommended to implement after the project registration. Project developers are required to maintain the monitoring report during the operation phase. Emission reductions are issued based on the monitoring report, which needs to be verified by the auditor (i.e. auditor #2). In CDM projects, the two auditors, i.e. auditor #1 and #2, should be different DOEs. In CDM projects, to avoid conflict of interest, validation and verification cannot be done by the same DOE, which may not be requested by other offsetting project standards.

The verification report is submitted to the CDM EB for certification and issuance of CERs. The issued CERs are then transferred to the CDM registry account of the relevant project participant after the mandatory fees are paid to the UNFCCC secretariat.

At the commercialization stage, there may have a few players including project developer, brokers, traders, exchange, wholesale providers and final buyers, depending on which channel the credits being sold. The companies who buy the carbon credits to fulfill their carbon reduction goals are normally the final buyers. In the next section, we will discuss what needs to be considered when choosing carbon credits.

3.5.2 How to Choose the Right Carbon Credits

Companies or individuals can purchase carbon credits directly from the project developer or offset providers. In the past, the author has conducted carbon consulting services for many companies in Hong Kong and other cities in Asia, including the listed companies, MNCs, SMEs, learned societies and NGOs. It is found that while companies have various objectives in carbon offsetting, majority of the organizations, even with their in-house sustainability team, have no ideas what to choose when purchasing the carbon credits. Business objective is always of paramount importance to all the business decisions within the organization. Company would buy carbon credits to offset its own carbon emission to reach its reduction target. It would also offer the offsetting programmes to its clients such as the green flights in the aviation industry or low-carbon accommodations in the hospitality sector. Some organization would also purchase the carbon credits and use them as souvenirs or gifts for the guests or the participants of the events organized for their own staff or their stakeholders. More examples will be discussed in Chap. 5.

Besides the business objectives, there are a few things companies should consider when they choose the carbon credits for offsetting purpose:

- Carbon offset standards
- Project types
- Project locations
- Sustainability benefits
- Value for money

The best carbon offsets are always those that best match the company's business philosophy and sustainability strategy, that could increase its reputation for corporate social responsibility and that could reinforce its brand name. Companies shall understand the different aspects of carbon offset standards and the differences among different standards, as summarized in Table 3.4. Carbon credits' prices would differ a lot based on different standards. Without the deep level of understanding, it is difficult for companies to plan ahead for the budget allocation.

According to company internal policy and/or sustainability guidelines, companies may exclude some types of projects, for example, one public organization in Hong Kong the author worked before would not consider the big hydropower projects for their very likely negative social impacts generated from these projects. Instead, companies may prefer a certain type of projects, for instance, many agricultural companies like methane capture projects from anaerobic digestion of manures, and construction and development companies may consider the afforestation project. Companies may also choose the credits from projects in some strategic partner countries or location.

Sustainability benefits to the local community of the project site shall be aligned with the company's sustainability goals and strategy. From the exercise of purchasing carbon credits, it is clear that carbon management shall be well embedded into the company's overall business strategy and involved in its business decision-making process. At the same time, any decision in choosing carbon credits would well reflect the company's goal and business and sustainability objectives. Bear in mind, purchasing carbon credits and carbon management as we will discuss soon are not a stand-alone and nice-to-have sustainability task by the sustainability team but shall be well addressed and engaged at all levels of the company, i.e. from the C-suit to the frontline staff.

References

Almendra F, West L, Zheng L, Forbes S (2011) CCS demonstration in developing countries: priorities for a financing mechanism for carbon dioxide capture and storage, World Resources Institute Working Paper

Carbon Disclosure Project (2018) The world's renewable energy cities. https://www.cdp.net/en/cities/world-renewable-energy-cities

Chan HR, Chupp AB, Cropper ML, Muller NZ (2017) The impact of trading on the costs and benefits of the Acid Rain Program, discussion paper from RFF's Regulatory Performance Initiative. www.rff.org/regulatoryperformance

David Suzuki Foundation (2009) The Problems with Offsets from Tree Planting Archived from the original on 2010-02-12

Dhillon A, Harnden T (2006) How Coldplay's green hopes died in the arid soil of India. The Telegraph, April 30 2006

Ecosystem Marketplace (2018) Voluntary carbon market insights: 2018 outlook and first-quarter trends. By Kelley Hamrisk and Melissa Gallant. June 2018

Figueres C, Philips M (2007) Scaling up demand-side energy efficiency improvements through programmatic CDM. Energy Sector Management Assistance Programme (ESMAP) technical paper series; no. 120/07. World Bank, Washington, D.C.

Hinostroza ML, Cheng C-C, Zhu X, Fenhann JV, Figueres C, Avendano F (2007) Potentials and barriers for end-use energy efficiency under programmatic CDM, C4CDM working paper series, vol 3. Risø National Laboratory, UNEP Risø Centre, Roskilde

IPCC (2018) Global Warming of 1.5°C, An IPCC Special Report on the impacts of global warming of 1.5C above pre-industrial levels and related global greenhouse gas emission pathways, in the context of strengthening the global response to the threat of climate change, sustainable development, as efforts to eradicate poverty

Kollmuss A, Zink H, Polycarp C (2008) Making sense of the voluntary carbon market: a comparison of carbon offset standards, WWF Germany

Qi S, Cheng S (2018) China's national emissions trading scheme: integrating cap, coverage and allocation. Clim Pol 18(Suppl 1):45–59

REN 21 (2018) Renewables 2018 global status report, REN21

Wolfgang S, Dienst C, Harmeling S, Schüwer D (2007) Renewable energy and the clean development mechanism, potential, barriers and ways froward, a guide for policy-makers. Federal Ministry for the Environment, Nature Conservation and Nuclear. Safety Public Relations Division. 11055 Berlin

World Bank and Ecofys (2018) State and trends of carbon pricing 2018 (May). World Bank, Washington, D.C. https://doi.org/10.1596/978-1-4648-1292-7

Chapter 4
Carbon Management Concepts

Carbon management is about understanding how and where an organization's activities generate greenhouse gas emissions, in order to then minimize these emissions in an ongoing and financially sustainable way. It extends from internal activities to the consumption of an organization's products or services and, ultimately, is about incorporating an understanding of carbon data into strategic business decision-making. There are four key points in understanding the definition of carbon management.

- Measurement of carbon footprint, through which to understand how and where emissions generated
- Mitigation measures to reduce the carbon footprint
- Cost-effectiveness
- Covering the whole value chain and embedded into business strategy

4.1 Business Drivers

Figure 4.1 summarizes the key business drivers for a company to conduct carbon management. We will discuss several drivers in details in the following section.

4.1.1 Cost Saving

Companies may have their own vision and ambition, but to increase the profitability or to reduce the cost is the key business objective to all business. As we discussed in Chap. 2, the example of Walker's Crisps tells us that measurement of carbon footprint could help to identify the opportunities throughout the operations and understand how to reduce the emission, while the operational boundary of the

© Springer Nature Switzerland AG 2020

S. W. W. Zhou, *Carbon Management for a Sustainable Environment*,

https://doi.org/10.1007/978-3-030-35062-8_4

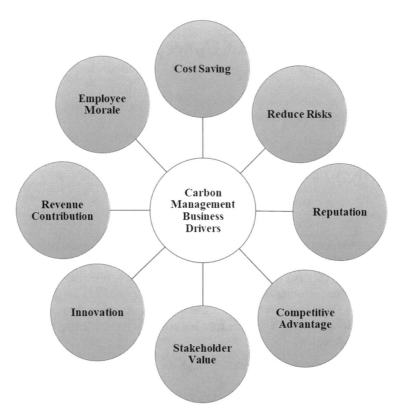

Fig. 4.1 Key business drivers for carbon management

carbon footprint measurement ensures us that carbon footprint is related to the activities such as electricity consumption, raw materials usage and waste disposal, which all have an operational cost to the reporting entity. Hence, to manage carbon with an objective to minimize it could offer a good opportunity for the company to check its activities and to find ways to reduce the emission, at the same time, reducing its operational costs. Back to Walkers' Crisps example, its 7% carbon reduction has saved the company more than £400,000 from 2007 to 2009, which Walkers invested in future energy-saving projects.

To better reflect this business driver, companies now capitalize the emission by putting an internal price to carbon. CDP.net reported that corporate use of an internal price on carbon is becoming the new normal for major multinationals, and in 2017 almost 1,400 companies were factoring an internal carbon price into their business plans, representing an eightfold leap over 4 years. Like Puma and Kering we discussed in Chap. 2, both have conducted an environmental financial statement that add resources, pollutions, waste and emissions into its balance sheet. Internal carbon price also provides an effective risk mitigation measures, which we will discuss in the next section.

4.1.2   Future Business Risks

Risk is the main cause of uncertainty in any organization. Companies must identify risks and manage them before they even affect the business. Enterprise risks include compliance or regulatory risk, financial risk, operational risk and reputational risk. Carbon footprint could be related to all these types of risks to an organization, while carbon management could provide a way to manage the relevant risks.

Table 4.1 summarizes the current carbon legislations or guidelines for different governments. With the increasing prominence and urgency of climate change, government will continue to introduce legislation and regulatory schemes to address climate change. Future legislation and regulations could include changes in government climate policy, local energy targets, the establishment of carbon trading schemes, carbon tax or tariffs, etc. By early adoption of carbon management, companies can evaluate their capability at an early stage to avoid these future climate-related risks, protect their brand reputation and begin to take advantage of carbon management-related business opportunities.

As we discussed, giving carbon an internal cost provides a way to deal with such future financial risk due to the coming climate-related legislation: to mitigate risk from current or expected regulation imposing a carbon price and as a financial tool in their risk management and investment processes (Topping et al. 2015). The TCFD recommends that companies should disclose the metrics and targets they are using to assess their corporate progress in managing climate-related risks and opportunities (TCFD 2017). BlackRock, the world's largest asset owner, who owns US$5.1

Table 4.1   Carbon legislations and guidelines

Government	Legislations or guidelines
Australia	The National Greenhouse and Energy Reporting Act 2007 (NGER Act) introduced a single national framework for reporting and disseminating company information about greenhouse gas emissions, energy production and energy consumption
EU	EU law on climate change and protection of the ozone layer include Greenhouse Gas Monitoring and Reporting, EU Emission Trading System, etc.
New Zealand	Zero Carbon Bill starting pubic consultation in October 2018
United Kingdom	The Climate Change Act 2008 to set a target for the year 2050 for the reduction target; to provide for a system of carbon budgeting; to confer powers to establish trading schemes for the purpose of limiting emissions or encouraging activities that reduce such emissions or remove greenhouse gas from the atmosphere, etc.
New York City	The legislation, outlined in the Blueprint for Efficiency, requires the city to cut energy in the biggest buildings by 20% by 2030 to keep on a path to ultimately reaching an 80 percent reduction target by 2050
North America	The Climate Registry, voluntary greenhouse gas emission reporting platform
Hong Kong	Carbon audit guidelines, Carbon Footprint Repository for listed Companies in Hong Kong, voluntary guidelines and reporting platform. Mandatory ESG Reporting for listed companies in Hong Kong, where carbon emission is one of the environmental KPIs needs to be reported and managed

trillion under management, equivalent to 4.3 % of the world's *GDP* announced in March 2017 that it would expect companies to provide assessments of how climate change would affect their business (reuter.com).

4.1.3 Stakeholders' Demand

Managing carbon footprint and reporting it are also the demands from different stakeholders, including government, NGOs, investors, clients, employees, etc.

Climate Action 100+ is an investor initiative to ensure the world's largest corporate greenhouse gas emitters take necessary action on climate change. More than 300 investors with over $32 trillion in assets collectively under management are engaging companies on improving governance, curbing emissions and strengthening climate-related financial disclosures. The companies include 100 "systemically important emitters", accounting for two-thirds of annual global industrial emissions, alongside more than 60 others with significant opportunity to drive the clean energy transition (climateaction100.org).

Consumers also have the same demand. An international study by Unilever revealed that more than 50% of people want to buy brands that are more sustainable and a third of consumers already purchase from brands they believe are doing social or environmental good (Unilever 2017). Another paper in the UK revealed that consumers are prepared to pay a price premium for carbon footprint labels but that label scepticism and fatigue as well as a lack in awareness about the impact of food production and consumption on climate change are major barriers for climate-friendly purchase behaviour (Feucht and Zander 2017). Saving cost by managing the carbon footprint and giving a carbon label to its product could give companies competitive advantages, when more and more customers are willing to buy low-carbon products.

Another very important stakeholder of the company is the employees. A study by Walsh and Sulkowski (2010) finds that working for an environmentally-friendly company is more important to employees than working for a financially successful one. Sustainability performance and its related communications could boost staff morale. Staff engagement is an important task for sustainability team, not only because behavioural change is a key way to achieve sustainability target, it also increases employees' satisfaction and employee retention. A staff engagement programme around sustainability issues such as energy saving, low-carbon education or community programmes shows that the company cares about both its employees and the environment, which could boost staff morale, increase company's reputation and attract more employees, especially the young generation.

4.1.4 Innovations

The late Prof. C.K. Prahalad pointed out that big social and environmental challenges present immense untapped market opportunities. He urged companies to create what he called "Next Practices", since incremental improvements to existing practices, or so-called best practices are simply not adequate (Prahalad 2005). What is innovation? "Next Practices" are all about innovation: imagining what the future will look like, identifying the mega-opportunities that will arise and building capabilities to capitalize on them.

Paris Agreement could then be a catalyst in driving innovation to design the green products, to improve business operations and processes to become more efficient, with a goal of carbon reduction. Low-carbon-driven innovation could increase both bottom line and top line. Researchers suggested adopting disruptive low-carbon innovation to trigger technological transition in developing countries (Wang and Chen 2008). Whereas sustaining innovations improve on the existing product or service attributes valued by end-users, disruptive innovations offer novel attributes, which could effectively create a new market, a new set of demands and preferences (Wilson 2017). Here we give a well-known example of Fuji Xerox's Integrated Recycling System and more cases on disruptive innovation such as shared mode transportation will be discussed in details in Chap. 7.

Case: Fuji Xerox Integrated Recycling System

Long before circular economy and innovation became the buzz word, Fuji Xerox has already been redefining its business model, from copy-machine products vendor to document service provider and developed its innovative Integrated Recycling System.

This system is composed of three concepts, with the main concept being the closed-loop system, where used products are effectively used as resources. The other two concepts include inverse manufacturing, which aims to create products under the premise of reusing parts in order to minimize environmental impact, and zero emissions, where parts that cannot be reused are separated and recycled to be utilized again as new materials.

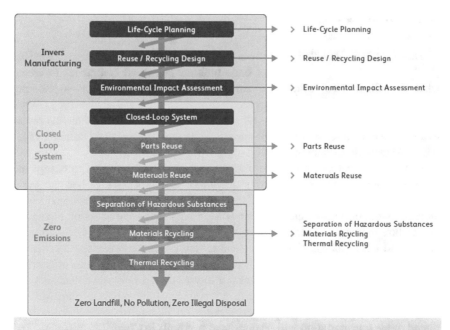

Building on this recycling system, the takeback program played a significant role in helping Fuji Xerox consistently achieve a recycling rate up to 99.5% for their end-of-life products at their business sites in Japan and the Asia Pacific region – an achievement unique to only Fuji Xerox in the document management and solution industry.

(Source: https://www.fujixerox.com/eng/company/ecology/cycle)

4.2 Carbon Management Framework

Carbon management framework provides a step-by-step approach for organizations when formulating carbon management strategic plan. It sets out actions and provides the basis for a robust carbon management strategy when applied in order. Based on its business nature or operation characteristics, organizations may select the different steps. But measurement, target setting and reduction are the essential steps, which should be included in carbon management plan for all organizations. The one with only measurement, target setting and reduction is called basic carbon management.

As the measurement, sequestration and carbon offsetting have been covered in Chaps. 2 and 3, respectively, and reduction solutions will be covered in Chap. 7, in the following section, we will focus to discuss how to set target, what the difference between avoidance and reduction is and how to switch to other energy sources (Fig. 4.2)

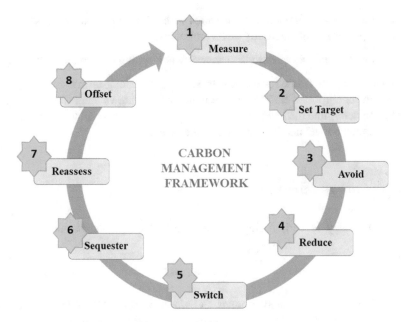

Fig. 4.2 Overview of carbon management framework

4.3   Target Setting

Setting targets is an important step, no matter for your personal development or for the business management process. Targets are different from objectives. For instance, based on ISO 14001:2004 definitions, "an objective" is an "overall environmental goal, consistent with the environmental policy, that an organization sets itself to achieve", and "a target" is a "detailed performance requirement, applicable to the organization or part thereof, that arises from the environmental objectives and that needs to be set and met in order to achieve those objectives".

The acronym SMART, with several slightly different variations, provides a comprehensive definition and an effective tool for target setting, which is also well applied in carbon management:

S – Specific
M – Measurable
A – Achievable, action-oriented
R – Realistic, relevant
T – Time-bound, trackable

For example, here is the EPA Victoria's Carbon Reduction Targets (EPA Victoria 2007):

– Be carbon neutral each year
– 10% reduction on our 2005–2006 energy-related GHG emissions by 2010

– At a minimum, implement any measures with a 4-year payback or less
– Green power for all electricity GHG emissions

In summary, specific questions to answer when setting carbon reduction target:

• What is the time frame for the targets, short-term or long-term?
• How to set corporate carbon reduction target (absolute vs. intensity)?
• How to determine the level of carbon reduction target (percentage of reduction)?
• How much budget needed and what is the payback?

4.3.1   Carbon Intensity

Carbon intensity is carbon footprint per unit of output or business measures (Table 4.2).

$$CI = CF / BM$$

where *CI* is carbon intensity (tonnes CO_2e/unit of *BM*); *CF* is absolute carbon footprint (tonnes CO_2e); *BM* is business measures of the reporting entity

Example 4.1
Given: China vows to reduce carbon intensity, the amount of CO_2 emitted per unit of *GDP*, by 40–45% from 2005 levels by 2020.

Question: How much will be China's absolute carbon footprint in 2020 (compared with that in 2005), given the average annual *GDP* growth rate is 8%?

Answer:

$$CI_{2020} = \frac{CF_{2020}}{GDP_{2020}} = \frac{CF_{2020}}{GDP_{2005} \times (1+8\%)^{15}}$$

$$CI_{2005} = \frac{CF_{2005}}{GDP_{2005}}$$

To achieve 40% reduction by 2020, based on 2005 level, it means

$$CI_{2020} \leq 60\% \, CI_{2005}$$

That is,

$$\frac{CF_{2020}}{GDP_{2005} \times (1+8\%)^{15}} \leq 60\% \frac{CF_{2005}}{GDP_{2005}}$$

Then,

$$CF_{2020} \leq 60\% \times (1+8\%)^{15} \times CF_{2005} = 1.90 CF_{2005}$$

It can be concluded that absolute carbon footprint in 2020 will be less than 1.9 times of that in 2005, which means the absolute carbon emission could still increase while the ambitious reduction target has been set in terms of carbon intensity.

4.3.2 Absolute Target vs. Intensity Target

Based on CDP, formerly Carbon Disclosure Project's 2011 report (CDP 2011), both absolute and intensity reduction targets are disclosed by S&P 500 respondents. There is no clear trend or preference when setting targets for different industries. For instance, in Information Technology sector, 14 companies use carbon intensity, 18 companies use absolute carbon and 6 use both. For Financials sector, 4 use carbon intensity and 21 use absolute carbon emission. But in CDP's 2014 report, CDP defined Climate Leaders are those setting absolute reduction targets, investing significant amounts in activities to achieve those targets and realizing tangible monetary savings and meaningful emissions reductions (CDP 2014). Table 4.3 gives out some examples of companies' reduction targets:

- Normalizing factors for intensity target could be revenue (e.g. GM, Microsoft), but more specific business measures may be used for specific industry, such as electricity generation for power company (e.g. CLP Group) and number of passengers or weight of the cargo for airport (HKIA).
- Companies started to set the absolute reduction targets in recently years (e.g. Microsoft, P &G).
- Some companies also looked into its supply chain or value, setting an absolute target to engage its suppliers and stakeholders.

Figure 4.3 shows the pros and cons of using absolute or intensity targets. Absolute targets could clearly assess the environmental impact and environmental cost in terms of GHG emission, while intensity target is easier to understand and to be used for comparison across a sector or a product category. But as our example #1 shows, intensity reduction target allows the emission to grow.

Table 4.2 Examples of business measures as the normalizing factor for carbon intensity

Reporting entity	Business measures or normalizing factor
Country/city	Gross domestic product (GDP)
Companies in general	Staff number; revenue
Manufacturer	Number of products
Property developer	Construction floor area (CFA)
Retail	Number of sales
Theme park	Number of visitors

Table 4.3 Examples of carbon reduction targets

Company	Absolute target	Intensity target	Normalizing factor	Target vs. baseline
CLP group		82%	Electricity generation	2050 vs. 2007
Hong Kong International Airport		10% with partners	Number of passenger or weight of cargo (in 100 kg)	2020 vs. 2015
General Motors		20% from facilities	Revenue	2020 vs. 2010
Microsoft	75% from operations	30%	Revenue	2012 vs. 2007 2030 vs. 2013
Procter & Gamble	30% from facilities	50%	Product unit	2012 vs. 2002 2020 vs. 2010
Coca-Cola	25% from value chain			2020 vs. 2010
Walmart	18% own operation 1 GT from supply chain			2025 vs. 2015 2025 vs. 2015

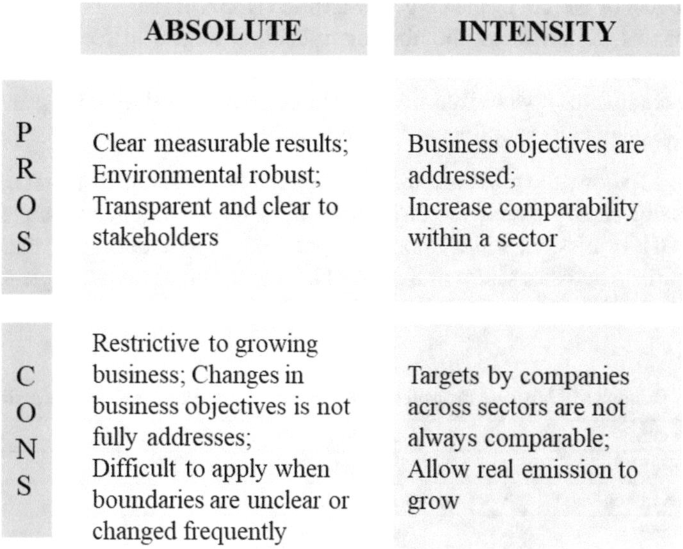

	ABSOLUTE	**INTENSITY**
P R O S	Clear measurable results; Environmental robust; Transparent and clear to stakeholders	Business objectives are addressed; Increase comparability within a sector
C O N S	Restrictive to growing business; Changes in business objectives is not fully addresses; Difficult to apply when boundaries are unclear or changed frequently	Targets by companies across sectors are not always comparable; Allow real emission to grow

Fig. 4.3 Comparison between absolute and intensity carbon reduction targets

4.3.3   Science-Based Targets

Science-based targets is a joint initiative of CDP, the UN Global Compact (UNGC), the World Resources Institute (WRI) and WWF. According to it, targets adopted by companies to reduce GHG emissions are considered "science-based" if they are in line with the level of decarbonization required to keep global temperature increase below 2 °C compared to preindustrial temperatures, as described in the Fifth Assessment Report of the Intergovernmental Panel on Climate Change (IPCC).

There are three science-based target (SBT) setting approaches:

- Sector-based approach: so-called the Sectoral Decarbonization Approach (SDA). The method is based on the 2°C scenario developed by the International Energy Agency (IEA) as part of its publication, Energy Technology Perspectives (ETP) 2014. The global carbon budget is divided by sector and then emission reductions are allocated to individual companies based on its sector's budget. The accompanying SDA tool allows companies to enter their data and determine their science-based targets according to the method.
- Absolute-based approach: The percent reduction in absolute emissions required by a given scenario is applied to all companies equally. When referring to this method at a global level, it is currently using the IPCC Fifth Assessment Report RCP2.6 subcategory that keeps overshoot to under 0.4 W/m^2, and which requires at least a 49% reduction by 2050 from 2010 levels to stay under 2C. This equates to at least a 1.23% absolute reduction per year.
- Economic-based approach: A carbon budget is equated to global *GDP* and a company's share of emissions is determined by its gross profit, since the sum of all companies' gross profits worldwide equate to global *GDP*. In this method, intensity targets would be considered science-based only if they lead to absolute reductions in line with climate science or are modelled using an approved sector pathway or method approved by the Science Based Targets initiative (e.g. the Sectoral Decarbonization Approach). Reductions must be at a minimum consistent with the low end of the range of emissions scenarios consistent with the 2 °C goal or aligned with the relevant sector reduction pathway within the Sectoral Decarbonization Approach.

Readers could get more detailed information, methods and tools for science-based target from its website: https://sciencebasedtargets.org/

4.3.4   Cost-Effectiveness

Another thing we need to consider when setting carbon reduction target is to understand how much cost or budget is needed to achieve the target. It is similar to energy audit, where consultants normally list out initial recommendations on energy saving measures with estimated energy and cost savings, investment cost and payback period for each measure. There are different methods to evaluate the cost and benefits to determine the most cost-effective option to achieve a certain carbon reduction. Three commonly used methods for cost-benefits analysis in carbon management are introduced here.

Life Cycle Cost Analysis (with Net Present Value)

According to ISO 15686-5:2017, life cycle cost is "the cost of an asset, or its parts throughout its life cycle, while fulfilling the performance requirements", and life cycle costing is "the methodology for systematic economic evaluation of life cycle costs over a period of analysis as defined in the agreed scope". Life cycle cost analysis takes into account all the relevant costs through a project's life cycles, including procurement, operating, maintenance, repairs and disposal. For example, for a building's life cycle cost may include the following:

- Construction costs (year zero costs)

  - Construction, fees, decanting etc.

- Maintenance costs and cycles

  - Major replacement, minor replacement, redecorations etc.

- Operation costs and cycles

  - Cleaning, energy, administration etc.

- Occupancy costs

  - Reception, catering, occupant's security etc.

- End of life costs

  - Disposal, reinstatement, continued value, etc.

All the costs are usually discounted and total to a present-day value known as net present value (NPV).

$$PV = \frac{A}{(1+i)^n}$$

where A = Future value, PV = Present value, i = discount rate, n = time.

Example 4.2 Life Cycle Cost (LCC)

Given: A new adsorption chiller costing $100,000 is installed in a CHP system for a building. It costs $20,000/year to operate over its 10-year life.

Question: What is the LCC at a 10% discount rate?

Answer:

LCC = PV (purchase cost) + PV (operating cost)

PV (purchase cost) = $100,000

PV (operating cost) = $20,000 [P/A, 10%, 10]$*$ = $20,000 × 6.145$*$

LCC = $100,000 + $20,000 × 6.145 = $222,900

*This can be found from compound interest table or be calculated by the following equation:

$$PV(i,N) = \sum_{t=0}^{N} \frac{R_t}{(1+i)^t}$$

In this example, R = cash flow = $20,000, i = discount rate = 10%, N = 10. A carbon reduction solution with a lower LCC is a more attractive option.

Savings-to-Investment Ratio (SIR)

Savings to investment is similar to return on investment (ROI). It takes the total cost savings over the lifetime of the measures divided by the upfront cost of the investment. This calculation may or may not include increases in energy prices or inflation rates.

$$SIR = \frac{Total\ present\ value\ of\ all\ Savings}{Investment}$$

The *SIR* ratio will give an estimate of the number of times a carbon reduction measure will pay for itself in savings, over its entire life time.

Example 4.3

Given: A club-house operator in Hong Kong engaged a consultant to conduct an energy audit for its various facilities. The consultant identified the following lighting upgrade opportunity during an on-site energy consultation.

Option	Exiting lighting fixture	Type for replacement	Capital cost (HK$)	Annual saving (HK$)
1	Halogen lamps	LED lamps	251,040	35,720
2	CFL downlight	LED downlight	526,800	21,500
3	T8	T5	923,780	200,450

Question: What is the SIR for different options and what is the favourable option?

Answer:

Assuming this company's minimum attractive rate of return is 10% and the new lighting system is expected to last 10 years, then SIR can be calculated as:

$$SIR = \frac{Annual\ savings \times [P/A,\ 10\%,\ 10]}{Capital\ cost}$$

SIR_1 = $35,720 × 6.145/$251,040 = 0.87
SIR_2 = $21,500 × 6.145/$526,800 = 0.25
SIR_3 = $200,450 × 6.145/$923,780 = 1.33

Hence, option 3 is an attractive option for lighting retrofitting in this case, as its SIR is greater than 1.0 and greater than other two options.

Simple Payback Period

Payback is simply the number of years it takes for the cost savings from a carbon reduction measures to equal the upfront cost. It answers how long before the investment pays for itself.

$$Simple\ Payback\left(years\right) = \frac{Cost}{Savings\ per\ year}$$

Modified payback could incorporate an estimate of increasing carbon prices, discount rates, and other factors to determine a more realistic value of the savings. Simple payback is commonly used as a selection criterion for reduction solution to achieve reduction target.

Taking the previous example, the calculated simple paybacks for three options are as follows:

Option	Exiting lighting fixture	Type for replacement	Capital cost (HK$)	Annual saving (HK$)	Payback (years)
1	Halogen lamps	LED lamps	251,040	35,720	7.0
2	CFL downlight	LED downlight	526,800	21,500	24.5
3	T8	T5	923,780	200,450	4.6

4.3.5   Social and Community Benefits

I have discussed in detail above how to set the SMART reduction target as the second step in carbon management, which includes how to differentiate the absolute reduction target and intensity target, how to set science-based targets to contribute the decarbonization of 2 °C pathways and how to set aside a budget for the reduction target and what would be the criteria for determining and evaluating the cost-effectiveness of the different options to achieve a certain target. Besides these, I would like to emphasize two points which I think are of paramount importance, but organizations normally overlook them when setting their targets.

Firstly, greenhouse gas emission reduction targets should be aligned with other community and social goals. In other words, organizations should focus on the society and community when setting the targets. Ontario Ministry of Environment and Climate Change set a very good model by publishing a guide on community reduction planning which focus on the engagement and inclusiveness as part of the main tools in planning emission reduction (Ontario Ministry of Environment and Climate Change 2017). Community programs such as engaging suppliers and customers could accelerate the reduction and organizations could set a more ambitious target based on these programs.

Secondly, besides the economic benefits, we should also assess the other benefits from carbon reduction. Co-benefits include:

- Creating additional jobs or employment shifted to low-carbon field
- Improving public health
- Stimulating innovation and trigger technology diffusion, adaptation and experimentation
- Providing social benefits to vulnerable populations

PricewaterhouseCoopers (PwC), one of the Big Four auditors, estimated the societal impact of its total greenhouse gas emissions at £65 m in 2017, and only 0.1% of it was attributable to their direct operations. The vast majority fell outside of its own operations. It concludes what we discuss here: reducing emission could have great social impact and companies should understand the emission along its value chain and work out reductions with its suppliers and other stakeholders.

4.4 Avoidance vs. Reduction

The most effective way to reduce the carbon impact is to avoid any activities that would produce direct GHGs or indirect energy-related emissions. Avoidance means there is completely no emission from that activity, while reduction means the emission from the same activity is lower. Table 4.4 lists the examples of avoidance and reduction for different activities.

As illustrated in Fig. 4.2, when planning carbon reduction measures, it is advised for organizations to prioritize the avoidance of emissions first (i.e. step 3), then their reduction through energy saving and energy efficiency measures (i.e. step 4) followed by the replacement of high-carbon energy sources with low- or zero-carbon alternatives (i.e. step 5). Companies should evaluate and prioritize the measures based on the cost-effectiveness criteria discussed above. Since avoidance measures normally have no upfront capital investment, it is treated as the most effective solution, which could not only reduce emission but also other environmental cost and energy cost. Hence companies should put effort in exploring and sourcing the technically available and feasible avoidance options first.

For overall reduction measures, a commonly used method is called GHG Marginal Abatement Cost Curve (MACC). Internationally there have been various

Table 4.4 Examples of avoidance and reduction solutions

Activity	Avoidance	Reduction
Energy use	Switch off electrical and electronic appliance when not in use; using natural lighting	Choose energy-saving appliance
Transportation	Walking, cycling	Public transportation
Products	Reusable food container, stainless straw	Reduced consumerism, recycling

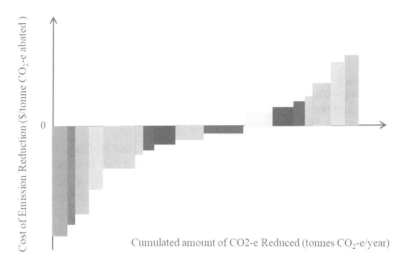

Fig. 4.4 GHG Marginal Abatement Cost Curve (MACC)

attempts to develop global MACCs (McKinsey and Company 2008, 2009; Faber et al. 2011) and MACCs for agriculture in particular (FAO 2012). While global MACCs are normally top-down analysis by using macroeconomic model, it could also be used bottom-up to model abatement potentials and costs for individual technologies and measure. Companies could also develop a company-specific GHG Marginal Abatement Cost Curve. As illustrated in Fig. 4.4, GHG MACC presents measures to reduce GHG emissions in order of their cost-effectiveness. The y axis is the costs of options available to a company to reduce its GHG emissions, and x axis is the accumulated reduction over the year. Each column represents one option with the height of the column its cost and the width of the column the reduction which could be achieved by implementing this option.

4.5 Switch

The next step on the carbon management framework is to switch to low-carbon-intensive energy source. This could include recover energy from waste or use cleaner energy source, like renewable energies.

4.5.1 Cogeneration and Trigeneration

Cogeneration is the simultaneous production of electricity and heat from a single fuel source. Combined Heat and Power or CHP is the most common application, where the exhaust heat produced during power production is recovered, for instance, by a heat exchanger, and can then be used to produce steam, hot water, hot air or

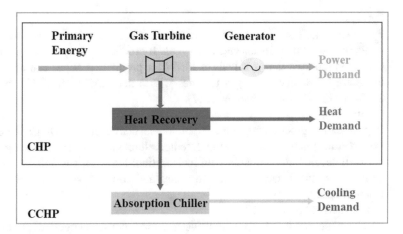

Fig. 4.5 Co- and trigeneration processes

another hot fluid. Combined Cooling and Power or CCP is another application where cooling is produced along with electricity from the recovered heat. Trigeneration, also called Combined Cooling, Heat and Power or CCHP, extends the cogeneration principle by coupling heat and power production with cooling production. As illustrated in Fig. 4.5, some of the heat produced by a cogeneration plant is used to generate chilled water for air conditioning or refrigeration. An absorption chiller is linked to the CHP to provide this functionality as an example.

Cogeneration is a promising solution in energy saving and carbon reduction. Compared with conventional power generation, where around 65–70% of energy will be lost in exhaust heat, CHP could achieve an overall efficiency of 85–90% with electrical efficiency of 30–35% and thermal efficiency 50–55% and produce only 10–15% loss. Currently cogeneration generates 11% of EU's electricity and 15% of its heat, reducing 200 million tonnes of CO_2e. Cogeneration is used in many key European industries, i.e. pulp and paper, alumina, chemicals, ceramics, glass, textiles, food and drink, as well as household (Cogen Europe 2018).

4.5.2 Waste-to-Energy (WtE)

Waste-to-Energy (WtE) or Energy-from-Waste (EfW) technologies consist of any waste treatment process that creates energy in the form of electricity, heat or transport fuels from a waste source. The most common application by far is incineration of the municipal solid waste (MSW) in a CHP plant. WtE technologies can be classified as follows:

- Thermal conversion

  - Incineration
    The combustion of organic material such as MWS with energy recovery is the most common WtE implementation.

– Pyrolysis
  Pyrolysis is an advanced thermal process. It takes place in the temperature range of 400–800 °C in absence of oxygen. It produces pyrolysis gas, oil and char. It is reported that pyrolysis of MSW for production of gas can be used for energy recovery using gas turbines with a net conversion efficiency of 28–30% (Kumar and Samadder 2017).
– Gasification
  In gasification process, organic compound gets converted into syngas in controlled atmosphere of oxygen at high temperature. Gasification technology is not well developed, compared with incineration; however, gasification could deliver a solution with higher efficiency and lower cost (World Energy Council 2013).
– Refuse-derived fuel (RDF)
  RDF is a fuel product from various wastes, and it consists largely of combustible components of such waste. The non-combustible fractions are separated by different processing steps, such as screening, air classification, ballistic separation, separation of ferrous and non-ferrous materials, glass, stones and other foreign materials and shredding into a uniform grain size, or also pelletized in order to produce a homogeneous material which can be used as substitute for fossil fuels. Its higher combustible and more homogenous energy content makes it attractive for energy production from waste (World Energy Council 2013).

• Biological conversion

– Anaerobic digestion (AD)
  Anaerobic digestion is a process of microbial degradation of organic biodegradable matter in absence of oxygen that produces biogas. It is commonly used at the wastewater treatment plant to stabilize the sludge. It has been also used in treating organic waste, such as manures and food waste. We will discuss in details the newly commissioned AD plant in Hong Kong in Chap. 7.
– Bioethanol production
  The main stages of a typical bioethanol production process include pretreatment, hydrolysis and fermentation. Lignocellulosic biomass is the most abundantly available raw material on the Earth for the production of bioethanol. Waste lignocellulosic biomass is mainly the agriculture waste such as sugarcane bagasse, straw and crop residues.

4.5.3 Renewables

As we discussed in Chap. 3, renewable energy is energy collected from renewable sources include solar, wind, geothermal, tidal, biomass, hydro, etc. As a user of energy, the organization has two options when switching to renewable energy sources: onsite installation of renewable projects or purchasing the energy produced from renewable sources. This induces two questions:

- What is the distributed energy generation option?
- How can companies purchase green energies?

We will discuss the distributed energy generation in Chap. 7. In this section, we talk how companies purchase renewable energies further.

According to the International Renewable Energy Agency (IRENA), companies follow two broad approaches in their consumption of renewable electricity:

- A passive approach in which consumption is based on the average renewable electricity content available in the grids from which companies source their electricity.
- An active approach in which consumption is based on actions companies take to procure or produce the renewable electricity they consume.

Corporate sourcing of renewables occurs when a company actively consumes, produces or invests in renewable energy to sustain its operations. IRENA defines four various models for corporate sourcing renewable electricity (IRENA 2018):

- Unbundled energy attribute certificates (EACs)
A company purchases attribute certificates of RE separately from its electricity. Examples of certificates are guarantees of origin (GOs) in Europe and renewable energy certificates (RECs) in the USA.
- Power purchase agreements (PPAs)
A company enters into a contract with an independent power producer, a utility or a financer and commits to purchasing a specific amount of RE at an agreed price and for an agreed period of time.
- Renewable energy offerings from utilities or electric suppliers
A company purchase RE from its utility either through green premium products or through a tailored RE contract, such as a green tariff programme.
- Production for self-consumption
A company invests in its own RE systems, on-site or off-site, to produce electricity primary for its own consumption.

In IRENA report (IRENA 2018), the overall volume of renewable electricity consumption from the reporting companies totalled 275 TWh in 2016, which represents 11% of their total electricity consumption. Among this 275 TWh, 61% of renewable electricity were sourced from procurement, while 39% were from self-generation.

Companies should develop their own renewable energy strategies, which could reflect companies' needs such as reaching their carbon targets and companies' mission or ambition to demonstrate their environmental stewardship. It would also take into account companies' constraints in operations in diverse geographic and regulatory settings. For instance, in Hong Kong, starting from 1 January 2019, both electric companies China Light Power and Hong Kong Electric launched RECs to their customers, but there is no option of PPAs at the moment. This is also the reason that a free electricity market is needed as the market instrument to offer options to users and offer opportunities in carbon reduction from alternative electricity generation (Hon Kong Consumer Council 2014).

4.6   Reassessment and Offsetting

What is the residual carbon emission after the implementation of all onsite reduction options or purchase of green power discussed above? Reassessment as the 7th step of the carbon management framework shown in Fig. 4.2 measures the residual carbon emission and assesses the current achievement, whether it reaches the reduction target set at step 2 and whether the solutions in activities avoidance, emission reduction, energy recovery from waste, use of renewable energies and natural or artificial carbon sequestration work as planned.

Reassessment is normally conducted in the same timeline as target setting. Intermittent assessment could also be conducted to check the performance, and that is better done by the online software tool as a real-time monitoring process. Reassessed emission level is used to compare with the baseline emission, which is obtained at the beginning of the carbon management process. The difference between the two is called residual carbon emission.

The last step of the carbon management framework is to offset the residual emission if any to meet the set target. Company should develop its own carbon offsetting policy and strategy. Based on what we discussed in Chap. 3, company shall consider the carbon offset standards, offsetting project types, project geographic locations, projects' sustainability or other co-benefits and the cost of the carbon credits.

4.7   Carbon Management Process

PDCA (plan–do–check–act) is an iterative four-step management method used in business for the control and continuous improvement of processes and products. It is also known as the Deming cycle/wheel (after its proponent, W Edwards Deming). As shown in Fig. 4.6, PDCA could also be adopted to describe the carbon management process:

- P – Strategic planning

    - Based on the carbon management framework, planning means to establish the objectives/targets and processes necessary to deliver results in accordance with the organization's business goals and carbon/sustainability policies.

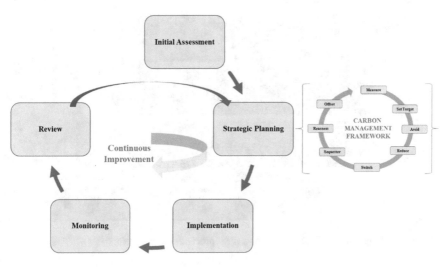

Fig. 4.6 Carbon management process

- It should be noted that when following carbon management framework, it is
  not necessary to cover all the steps, but company could select different steps
  in the process, based on company's needs and actual situation and constraints.
  Nevertheless, total carbon management covers all the steps till carbon
  neutrality, which as a typical carbon management process that will be further
  discussed in the Chap. 5.

- D – Implementation
  - Implement all the processes.

- C – Monitoring
  - Monitor and measure processes against policies, objectives, targets, legal and
    other requirements, and report the results.

- A – Review
  - Take actions to continually improve the carbon management process and
    carbon/sustainability performance. This includes the review of the whole
    process and system, whether they are up and running as planned; the correction
    if anything wrong; and the process upgrading as organizational experience
    acquired, which leads to the evolving organizational concept and back to the
    new planning stage with updated targets, policies and measures, etc.

As also illustrated in Fig. 4.6, initial assessment is separate from or prior to the
PDCA cycle. It is the part of initial preparation of the carbon management process,
during which company prepare to get started with the carbon management process.

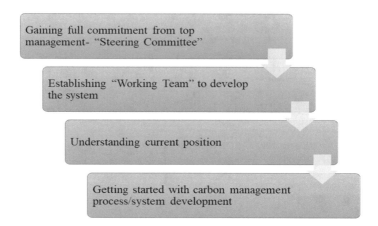

Fig. 4.7 Getting started with carbon management process/system development

4.7.1   How to Get Started?

Carbon management is a whole organization philosophy and concept, and it is applied across the organization's activities. It is a formally structured management approach to meet desired carbon target and tackle ever-increasing climate problems and impacts. The well-defined carbon management framework is a proactive solution to be implemented in the most cost-effective way. Figure 4.7 illustrates a few steps to get started: getting the full commitment from the top management, setting up a working team and understanding the current situation before starting with the carbon management framework and system development.

4.7.2   Top Management Commitment

The GlobeScan/SustainAbility Leaders Survey is the longest-running survey that has tracked expert opinions on sustainable development leadership for over 20 years. According to GlobeScan's 2018 report (GlobeScan and SustainAbility 2018), integrated sustainability values, sustainability as part of core business model/ strategic approach, and executive leadership with strong sustainability values/strong leadership are the top three aspects to define corporate sustainability leaders.

Companies could establish a "steering committee" for carbon/sustainability management as the first step to demonstrate leadership and commitment by leading from the top and wherever possible integration of carbon management into business processes. The role of top management, or steering committee as defined is to develop and approve carbon/sustainability policy as a statement of commitment; to

Table 4.5 Integrating sustainability into existing board committee responsibilities

Board committee	Link to sustainability
Corporate Governance Committee	Oversee matters of corporate governance, corporate responsibility, sustainability trends and ESG impacts to the business Review the director orientation and education program for ensuring the appropriate expertise and knowledge is present overall Assist in monitoring and reviewing corporate governance and reputational risk exposures Review company-wide policies regarding Corporate Governance Principles
Audit Committee	Promote listing standards for stock exchanges that include requirements for sustainability disclosure and have the responsibility of carrying out an ESG materiality assessment, as well as the outcome of the assessment within the annual financial filings Understand any risks and opportunities related with reporting on the sustainability performance of the firm Ensure the quality of communications and data around sustainability Ensure compliance with new regulations around sustainability
Compensation Committee	Link sustainability issues material to the business to ESG targets Integrate sustainability executive performance evaluations and compensation Identify any risks related to the compensation structures and the time horizon of the associated targets Engage with investors and stockholders around their proposals related to compensation matters
Nominations Committee	Integrate sustainability into the director nominations process Maintain an up-to-date skills record for board directors. Introduce sustainability skills in the directors' selection process Integrate sustainability into board performance evaluations Proactively communicate the succession plan

provide resources and support during the process; and to regularly review the process to ensure its continuing suitability, adequacy and effectiveness.

Instead of setting up a standalone sustainability steering committee, it is a better and more effective practice to integrating sustainability governance into corporate governance. The UN Environment Programme Finance Initiative (UNEP FI 2014) proposed a new model for sustainability governance and encouraged corporate to integrate sustainability into their existing board committees, as shown in Table 4.5.

4.7.3   Establishing Working Teams

Besides the top management commitment, company shall establish the working teams, where different roles, responsibilities and authorities and resources shall be well defined. The working teams could be at different levels, from senior executive level, to middle management and then cascade down to the staff level, with the purpose to assign the accountability and as well as to engage majority of the staff within the operations.

Unilever: Management Structures for sustainable Business Integrated into the Organizational Framework

Board level – governance and oversight	The Boards of Unilever (Unilever PLC and N.V.)
	   The Boards have the ultimate responsibility for the management, general affairs, direction, performance and long-term success of Unilever
	Corporate Responsibility Committee
	   Committee of Non-executive Directors oversees Unilever's conduct and governance of conduct as a responsible company
	   Monitors the progress – and potential risk – of the Unilever Sustainable Living Plan (USLP)
	Audit Committee
	   Oversees Unilever's approach to risk management, including the corporate risks and related mitigation/response plans
	   Reviews significant breaches of the Code of Business Principles as part of its remit to review risk management
	   Oversees the independent assurance programme for the USLP

Unilever Leadership Executive (ULE) – operational leadership of the business	The ULE, led by Chief Executive Officer is responsible for managing profit and loss and delivering growth. It monitors implementation and delivery of the USLP At ULE level, sustainability and corporate responsibility are championed and led by Chief Marketing & Communications Officer

Divisional and function leadership – strategic development of our goals	Unilever Sustainable Living Plan Steering Team
	   Supports the ULE and is accountable for ensuring our sustainability goals are achieved. It meets five times a year
	Foods and Refreshment, Beauty & Personal and Home Care Divisions
	   Responsible for the strategic development of our sustainability goals
	Chief Sustainability Officer
	   Leads the integration of sustainability into business and communications, working with business functions, divisions, brands and country teams to open up opportunities for business growth
	   Leads our external sustainability agenda to drive transformational change across markets through advocacy and partnerships

Source: www.unilever.com/sustainable-living/our-strategy/
our-sustainability-governance/

Fig. 4.8 Structure of Unilever Leadership Executive (ULE)

Taking an example of Unilever as in above case, Unilever Leadership Executive (ULE), the operational level leadership team, chaired by Chief Executive Officer, includes 11 members: Chief Financial Officer, Chief Digital & Marketing Officer, Chief Supply Chain Officer, Chief R&D Officer, Chief HR Officer, Chief Legal Officer, Chief Operating Officer, as well as President of Foods & Refreshment, President of Beauty & Personal Care, President of South Asia and President of Home Care, as shown in Fig. 4.8.

The following could be considered when setting up carbon management working team:

– What are the company's emission sources and who are the data owners of emission data?
– What is the business nature and how many business functions, operations or units within the company?
– Who are the business operation heads?
– Are property management, facility management and finance team be engaged?

4.7.4 Initial Assessment

Baseline assessment is to identify the organization's current strengths and weakness in tackling climate change issues. Many exercises can be done for initial assessment, such as initial review, gap analysis, CSR or carbon health check and baseline measurement or carbon audit.

Initial review or carbon health is an essential exercise aimed at understanding the organization's current position, such as its existing environmental programmes, systems and major environmental aspects including emission sources; it also checks the company's climate readiness, staff's overall awareness and the baseline performance, if the carbon audit has not been conducted yet. The approaches of initial include:

– Analysis of management structure and hierarchy
– Divisional and departmental analysis of activity
– Task analysis of activity
– Longitudinal and cross-sectional analysis of procedures and working instructions

Initial review is best conducted through discussions with different functions in the company, especially line managers and line workers, who have direct knowledge of as many relevant functions as possible. Information can normally be collected by a number of methods:

– Checklists
– Questionnaires
– Consultations
– Interviews
– Inspection and measurement

When conducting gap analysis, besides reviewing company's current performance, practices and emission profiles, consultants or corporate sustainability personal shall also understand the current or future risks the company will face, such as the company's environmental accident, fines or penalties, the existing or future government carbon and climate-related legislation and regulatory requirement or incentives, competitor's carbon performance or existing or coming sectoral standards, etc.

Baseline carbon emission measurement are also planned to be conducted in this stage. As discussed in Chap. 2, sessions 2.5.1 and 2.5.2, base year cannot be changed once defined, while baseline emission could be updated if there were significant changes in reporting entity's organizational boundaries or operational boundaries or others.

4.7.5   Summary of Tasks and Deliverables

As a carbon consultant or a sustainability manager, what needs to be delivered in the whole carbon management process? Table 4.6 summarizes the potential tasks or deliverables at different carbon management stages.

4.8   Case Study

Ocean Park, opened in 1977, is a marine mammal park, oceanarium, animal theme park and amusement park. It is Hong Kong's iconic theme park and currently is the second largest in Hong Kong, following Hong Kong Disneyland.

In 2010–2011, when I was working at Carbon Care Asia Ltd., we have been commissioned by Ocean Park Corporation (OPC) as an independent carbon management consultant to assist in conducting Park-wide carbon audits and developing a 10-year Carbon Footprint Management Plan. In this section, I will use OPC as a case to discuss briefly how carbon management framework and process could be adopted and bring benefits to the organizations. More information can be found from: /www.oceanpark.com.hk/en.

Table 4.6 What to deliver during carbon management process

Stage	What can be delivered
Initial assessment	Environmental review CSR assessment Carbon health-check Baseline measurement Carbon audit Facility (e.g. factory), process (e.g. production line) and product (e.g., carbon label)

Strategic planning	Develop carbon management policies and procedures and set reduction targets from the base line
	Reduction and offsetting
	   Energy efficiency solutions
	   Renewable energy installation (Green power procurement)
	   Carbon offsetting

Monitoring	Software and other real-time reporting platforms
Review	Reporting Verification Corrections and recommendations

Initial Assessment and Strategic Planning

In early 2010, Carbon Care Asia was commissioned by OPC to conduct the first carbon audit and strategic carbon planning, including to:

- Conduct an in-depth carbon audit and analysis in report format through site inspection, data assessment and interviews of key personnel of the Park
- Provide practical recommendations that demonstrate quantifiable, verifiable greenhouse gas reductions and on-site good practice in energy efficiency and sustainable energy in report format
- Render assistance to the management of the Park to devise strategic planning to create action pathways and to set a leading benchmark standard through the facilitation of a 2-day workshop
- Create educational programs and materials on carbon management knowledge with value creation for actions in the 2-day workshop
- Devise an effective media and communication strategy based on environmental efforts from existing and new implementations to be presented to the Park

The first carbon audit was conducted for 2008/2009, with a full emission profile covering Scope 1, 2 and 3. This first carbon audit data did not serve as the baseline during the target setting discussion, but this provided a full picture of the company's carbon emission, the sources, the magnitude and clearly showed the senior management at the strategic planning workshop how and where emissions came from and how much. In other words, this initial assessment project has paved the way to establish the baseline for OPC, to develop its carbon reduction plan and to chart its longer-term strategy to embrace low-carbon values.

Strategic Planning Project was conducted between March and June 2010, through scientific carbon measurement and rigorous assessment of carbon capacity and values via interviews and facilitated discussion with top management of OPC. Two rounds of questionnaires were circulated among up to a total of 22 senior executives at OPC. Interviews were conducted with 16 executives from Chief Executive, Executive Directors, to Senior Managers to collect their views and knowledge.

It was found during this stage that:

- OPC's broad support for Carbon Leadership initiative and the possibility of creating a special status with stakeholders by virtue of the strategy.
- No current objectives on carbon reduction.
- The environmental/carbon policy is not yet articulated.
- There is varied awareness of carbon leadership among different members of the leadership team: the key is to arrive at the lowest common denominator of understanding to begin with and progress with continuous knowledge building.
- Top leadership of OPC will determine the Carbon Strategy and allocation of resources for delivering the strategy.
- Tremendous branding and image perception opportunity exists. OPC could take a leading role in education and probably wants to be a LEADER in the carbon field.
- Quick wins/immediate benefits are needed to get employees and management based in the same mindset.

The above also became the deliverables for OPC, including the governance structure setup, the target setting, the 10-year roadmap and the policy development.

Governance and Engagement

OPC has the potential and capabilities to become a leading climate friendly "edutainment" business in the world. It can act as a Carbon Leader in the world by:

- Attracting attention to climate change response
- Nurturing interests and awareness of low-carbon culture among youths and wider public
- Inspiring new ideas and ways among young populace of low-carbon solutions

An Environment and Carbon Management Steering Committee, comprised of senior representatives from each division and a newly appointed carbon management officer, was established to set measurable objectives, to develop and to maintain environmental management and carbon reduction program.

The management of OPC is committed to good environmental practice and carbon care initiatives among its staff and business partners and will seek to ensure that suppliers and contractors of Ocean Park apply the same level of commitment and concern to environmental matters and carbon care when dealing with the business.

In February 2012, OPC invited several renowned green groups and like-minded corporations to share their experiences in carbon management, thereby helping OPC plan a series of carbon management programmes and set specific carbon reduction targets. These programmes and targets together constitute Ocean Park's Carbon Footprint Management Plan, which was approved by our Board of Directors in June 2012.

Target Setting
As a consultant, we did a very comprehensive review and present the outlined framework for OPC for its target setting. Firstly, we reviewed protocols, references for setting corporate carbon goals and reduction targets on the national and industrial levels, which includes:

- Kyoto Protocol
- Cancun climate agreement (UNFCCC COP 16 took place from 29 November to 10 December 2010 in Cancun, Mexico)
- National reduction targets for 2015, 2020 – EU, USA, Canada, Australia, China, HK
- Corporate reduction targets for 2015, 2020 – HKIA, Disney, theme parks, hotels
We then formulated the pathways of setting reduction targets, with examples and illustrations, especially discussed "absolute reduction", "carbon intensity" and "how to present the whole story". And lastly, we recommended options for OPC reduction targets for 2015, 2020, covering:
- Selected base year and baseline measurement
- Absolute emission and reduction
- Selected normalizing factor for carbon intensity and intensity reduction target
- Selected sites, products, programmes, activities, etc.
- How to present the whole story

With 2011/2012 chosen as the base year, Ocean Park is targeting to reduce its absolute carbon emissions by 10% and its carbon intensity (with the number of guests as the normalising factor) by 25%. These targets represent new standards in carbon management for the theme park industry.

Among the many carbon-reducing measures are the development of a highly energy-efficient Life Support System (LSS) for animals, the installation of solar-panels on electric carts and conversion of food waste to animal feed.

10-Year Roadmap
Besides the reduction in electricity consumption, OPC's Carbon Footprint Management Plan also covers the following areas:

- Putting in place staff awareness training programmes and encouraging them to adopt a low-carbon lifestyle at work and in their personal life
- Educating the public about the impact of climate change on polar animals and the planet through in-Park low-carbon experiences, educational courses and displays, with the aim of raising awareness of sustainable development among the public, thereby encouraging them to adopt a low-carbon lifestyle
- Conducting periodic progress reviews and monitoring carbon reduction achieved

Investing an estimated amount of over HK$50 million on carbon reduction programmes is expected to generate HK$70 million utility cost savings over the next 10 years. The Carbon Reduction Plan will be updated continuously as new initiatives are identified and adopted.

Environmental and Carbon Management Policy

Finally, all these have been articulated and formed into OPC's Environmental and Carbon Management Policy including its commitment to environment and community, its programmes and targets, and its governance. Below lists a few items from its policy:

– Ocean Park recognizes its corporate responsibilities towards both the environment and the community and is committed to the dual ethos of responsible global citizenship and sustainability in order to reduce carbon emissions and therefore, address climate change.
– Ocean Park will endeavour to ensure that the management of its operations is in a manner which adheres to environmental and carbon good practices.
– Ocean Park will use its knowledge base to help promote environmental well-being and carbon caring initiatives, by interacting positively with stakeholders in local communities, industry, commerce and the public sector and with the wider national and international communities
– Ocean Park recognizes direct and indirect environmental impacts and endeavours to manage these in a responsible manner.
– Ocean Park has developed a carbon reduction plan from 2011/2012 to 2021/2022. Ultimately, Ocean Park aspires to become carbon neutral. To this end, OPC will assess all new activities and projects for environmental impact and carbon emissions.

References

CDP (2011) CDP S&P 500 report 2011 strategic advantage through climate change action
CDP (2014) CDP S&P 500 climate change report 2014 climate action and profitability
Cogen Europe (2018) COGEN Europe response inception roadmap on "strategy for long-term EU greenhouse gas emissions reductions". Brussels, 10 August 2018
EPA Victoria (2007) Discussion paper draft carbon management principles
Faber J, Behrends B, Nelissen D (2011) Analysis of GHG marginal abatement cost curves. Delft, CE Delft, March 2011. Publication number: 11.7410.21.
FAO (2012) Using marginal abatement cost curves to realize the economic appraisal of climate smart agriculture policy options. In: The Ex ante appraisal carbon-balance tool. Food and Agriculture Organization of the United Nations
Feucht Y, Zander K (2017) Consumers' willingness to pay for climate-friendly food in european countries. International Journal on Food System Dynamics. Proceedings in System Dynamics and Innovation in Food Networks. https://doi.org/10.18461/pfsd.2017.1738
GlobeScan and SustainAbility (2018) The 2018 GlobeScan-SustainAbility leaders survey
Hong Kong Consumer Council (2014) Searching for new directions - a study of Hong Kong electricity market, December 2014
IRENA (2018) Corporate sourcing of renewables: market and industry trends – REmade index 2018. International Renewable Energy Agency, Abu Dhabi
Kumar A, Samadder SR (2017) A review on technological options of waste to energy for effective management of municipal solid waste. Waste Manag 69(2017):407–422
McKinsey & Company (2008) An Australian cost curve for greenhouse gas reduction

McKinsey & Company (2009) Pathways to a low-carbon economy – global greenhouse gases (GHG) abatement cost curve. Version 2 of the Global Greenhouse Gas Abatement Cost Curve.

Ontario Ministry of Environment and Climate Change (2017) Community emissions reduction planning: A guide for municipalities

Prahalad CK (2005) The fortune at the bottom of the pyramid. Pearson Education, Inc. Publishing as Wharton School Publishing

TCFD (2017) Final report: recommendation of the task-force on climate-related financial disclosure

Topping N, Cushing H, Law S, Pierce L (2015) Carbon pricing pathways toolkit: navigating the path to 2°C. CDP and the We Mean Business Coalition. September 24, 2015

UNEP FI (2014) Integrated governance, a new model of governance for sustainability. A report by the Asset Management Working Group of the United Nations Environment Programme Finance Initiative, June, p 2014

Unilever (2017) Making purpose pay: inspiring sustainable living

Walsh C, Sulkowski A (2010) A greener company makes for happier employees more so than does a more valuable one: a regression analysis of employee satisfaction, perceived environmental performance and firm financial value. Int Environ Rev 2010(11):274–282

Wang ZW, Chen J (2008) Achieving low-carbon economy by disruptive innovation in China. 2008 4th IEEE International Conference on Management of Innovation and Technology

Wilson C (2017) Disruptive low carbon innovations. ECEEE Summer Study Proceedings, pp 2149–2158

World Energy Council (2013) World energy resources: Waste to energy

Chapter 5
Total Carbon Management Model

Based on the carbon management concept discussed in Chap. 4, this book aims to encourage businesses to focus on their mitigation efforts, that is, companies should focus on reducing their GHG emissions, which I think is vital to company's climate strategy and also critical to globally collective effort in tackling climate impacts, more specifically, to achieve zero carbon emission by mid of this century.

Following carbon management framework (as shown in Fig. 4.2), company could select different steps in the framework. In the previous consulting job at Carbon Care Asia, we defined the basic carbon management including carbon measurement, target setting and reduction, while "total carbon management" including measurement, reduction and offsetting, which we called as "Mr. O". "Total carbon management" is still a consultancy solution offered by Carbon Care Asia now; details of which could be found from its website: https://www. carboncareasia.com/eng/Carbon_Solutions/Carbon_Management.php.

In this book, I will discuss my "upgraded" version of the above, and it is called Total Carbon Management Model.

In Chap. 1, I have introduced climate change risk framework and management process (Fig. 1.8) and emphasized that adaptation and mitigation are two complementary climate strategies for company to reduce and manage the climate risks. When the impacts of climate change are increasingly felt and discussed around the world especially after Paris climate conference, it has become even clearer that integrating adaptation and mitigation is essential to a company to increase its adaptive capacity and build its climate resilience and at the same time lower its carbon emission.

In this chapter, I will first introduce the concept of carbon neutrality and how to achieve it mainly built on the carbon management framework on mitigation. I will then introduce a few cases of carbon management in some sectors and who have achieved carbon neutral and elaborate some controversial cases and especially the mis-claims of carbon neutrality when not following the standards. Next, I will introduce how businesses could build their climate resilience and show the

© Springer Nature Switzerland AG 2020
S. W. W. Zhou, *Carbon Management for a Sustainable Environment*,
https://doi.org/10.1007/978-3-030-35062-8_5

interaction between adaption and mitigation. Finally, I will present the total carbon management model.

5.1   Carbon Neutrality

Carbon neutral or carbon-neutral became a buzzword in the beginning of this century, and carbon-neutral was selected as Word of the Year by Oxford Dictionaries US in 2006.

You may also have heard zero carbon, for instance, net zero carbon building, as defined by World Green Building Council (WorldGBC), is a highly energy-efficient building with all remaining operational energy use from renewable energy, preferably on-site but also off-site production, to achieve net zero carbon emissions annually in operation (WorldGBC 2017).

You may also have heard climate neutral. Climate Neutral Now is an initiative launched by the UN Climate Change in 2015, aiming at encouraging and supporting all levels of society to take climate action to achieve a climate neutral world by mid-century, as enshrined in the Paris Agreement adopted the same year. The initiative invites companies, organizations, governments and citizens to work towards climate neutrality by a simple three-step method: measure their GHG emissions, reduce them as much as possible and compensate those which cannot be avoided by using UN certified emission reductions (CERs).

So, what is the definition of carbon neutral?

5.1.1   Definition

There are different carbon neutral standards in the world, as listed below:

- Australia's Carbon Neutral Program: is a voluntary scheme which certifies products, operations or events as carbon neutral against the Australian Government's National Carbon Offset Standard (NCOS). This standard builds upon existing standards through its guidance on voluntary offsets and its minimum requirements for calculating, auditing and offsetting a carbon footprint to achieve carbon neutrality.
- Carbon Footprint Standard: developed by Carbon Footprint Ltd., a private company and a leading provider of carbon consultancy and service since 2004. To be qualified for the Carbon Footprint Standard Carbon Neutral, an organization, product, service or event must have first achieved the Carbon Footprint Standard CO_2e Assessed label. After that all emissions must be offset using carbon credits that have been generated by projects that meet the requirements outlined in the standard (Carbon Footprint Ltd. 2017).

- PAS 2060: A British standard but internationally widely used. Publicly Available Specification 2060 for the demonstration of carbon neutrality was first published in 2010 and revised in 2014. PAS 2060 specifies requirements to be met by any entity seeking to demonstrate carbon neutrality through the quantification, reduction and offsetting of GHG emissions from a uniquely identified subject (BSI 2014).
- The CarbonNeutral Protocol: created and managed by Natural Capital Partners and the first to provide a clear set of guidelines for businesses to achieve carbon neutrality back in 2002. It is revised every year in accordance with the latest business, scientific and environmental standards (Natural Capital Partners 2019).

The definitions of carbon neutral and/or carbon neutrality by different standards are summarized in Table 5.1.

In summary, carbon neutral is defined as the condition in which the carbon emission of the subject is net zero for a given period of time. Carbon neutrality can be achieved by measuring and reducing carbon footprint of the subject and offsetting the residual emission to reach net zero carbon footprint.

5.1.2 Steps to Demonstrate Carbon Neutrality

Figure 5.1 summarizes the basic steps to demonstrate carbon neutrality. Elaborations to key important points to be considered will be discussed, mainly based on PAS 2060. A summary of differences among different carbon neutral standards is shown in Table 5.2.

Example 5.1 Definition of Carbon Neutral
Given: Plant A had a production line #1, and carbon footprint of plant A in 2017 was 10,000 tonnes. Plant A set up a new production line #2 in 2018, and the carbon footprint of plant A in 2018 was also 10,000 tonnes.
  Question: Which subject from the above was carbon neutral?

(a) Plant A in 2018.
(b) Production line#1 in 2018.
(c) Production line#2 in 2018.
(d) None of the above subject was carbon neutral.

  Answer:

(a) Plant A in 2018 was not carbon neutral, as its emission was 10,000 tonnes.
(b) Production line #1 in 2018 could be carbon neutral if all 10,000 tonnes of emission were from other entities of plant A; otherwise, it could not be carbon neutral.

(c) Production line #2 in 2018 could be carbon neutral if 10,000 tonnes of emission were from other entities of plant A or, adding production line#2 had no contribution to the carbon emission of plant A; otherwise, it could not be carbon neutral.

(d) None of the above was carbon neutral in 2018, as long as the subject had emission, e.g. in 2018, the emission from production line#1 could be 5000 (reduced from 8000 tonnes due to reduction measures implemented), while the emission from production line#2 was 1500 tonnes, and the emissions from the rest of plant A was 500 tonnes.

Determine the Subject

According to PAS 2060, all subjects or entities could be applicable to demonstrate their carbon neutrality, which include but are not limited to:

- Activities.
- Products.
- Services.
- Buildings.
- Projects and major developments.
- Towns and cities.
- Events.

Australia's National Carbon Offset Standard (NCOS) only covers five categories, i.e. organizations, products and services, events, precincts and buildings. NCOS has separate standards for each category.

Table 5.1 Definition of carbon neutral by different standards

Standard	Definitions
Australia's National Carbon Offset Standard Carbon Neutral Program	Carbon neutrality: Refers to a situation where the net emissions associated with an organization's activities, product, services or events are equal to zero because the organization has reduced its emissions and acquired and cancelled carbon offsets for its remaining emissions
Carbon Footprint Standard	Carbon neutral: Having a net zero carbon footprint by balancing emissions released with an equivalent amount sequestered, avoided or offset
PAS 2060	Carbon neutral: Condition in which there is no net increase in the global emission of greenhouse gases to the atmosphere as a result of the greenhouse gas emissions associated with the subject Carbon neutrality – State of being carbon neutral
The CarbonNeutral Protocol	Carbon neutral: Condition in which the net GHG emissions associated with an entity, product or activity are zero for a defined duration

Table 5.2 Comparison of different standards in how to demonstrate carbon neutrality

Steps	Australia NCOS	Carbon Footprint Standard	PAS 2060	The CarbonNeutral Protocol
Determine the subject	Cover five categories, i.e. organizations, products and services, events, precincts and buildings; NCOS has separate standards for each category	Apply to organizations, services, products and events	Apply to all subjects, but need to define purposes, objectives and functionality of the subject	The subject is the entity, product or activity
Quantify CF	Paper use must be included in scope 3; for product, full life cycle must be assessed, unless justified and communicated. Organization must set a base year and base year re-calculation policy to allow for their emissions to be accurately compared over time	For product and service, the boundary definition should as a minimum include cradle-to-factory gate	For product, adopt PAS2050; include any scope 1, 2 or 3 emission estimated to be more than 1% of the total carbon footprint unless not technically feasible or cost-effective	Third-party deliveries of company goods must be included for organizations; for entities, assessments must be annual and relate to a 12-month period
CFMP	Emission management plan is required to identify reduction strategy over a specific timeframe; offsetting strategy. Must re-calculate the emissions baseline and update the emissions management plan over time	Not required	CFMP is required to state the commitment of carbon neutral, the timescale, the target setting, the reduction programmes and offsetting strategy. Updated in a 12-month time	Company should develop a GHG reduction plan to deliver internal emission reductions, taking into consideration the main sources of GHGs from the subject and the likely cost-effectiveness of alternative emission reduction actions

(continued)

Table 5.2 (continued)

Steps	Australia NCOS	Carbon Footprint Standard	PAS 2060	The CarbonNeutral Protocol
Reduce	No specific requirement on reduction target	Reducing absolute emissions; reducing emissions by per unit of turnover. Reductions against turnover should also be corrected for inflation; reducing emissions by per unit of production. Emission tracked over two or more years	Must monitor the subject's emissions intensity to a baseline; must implement CFMP to reduce the subject's emissions; emission reductions must be achieved and identified; mandatory targets would be regarded as the minimum reduction target, and outsourcing could not regarded as a form of reduction	Must commit to an overall net zero emission target and set an internal reduction target to ensure actual emissions decrease over time; the target may be expressed as an absolute GHG emission reduction or as a decrease in GHG intensity. No specific requirement on reduction target
Offset	May not use long-term OR temporary CERs	Follow Quality Assurance Standard (QAS) for Carbon Offsetting. Excluded offsetting projects: Large-scale Hydropower projects (>20 MW) Industrial gas projects (HFC-23 or N2O from adipic acid production) Methodologies Forestry-related unless based on sustainable REDD+ project	Adopted standards/ carbon credits: Kyoto Protocol's Clean Development Mechanism/CERs Kyoto Protocol's Joint Implementation/ ERUs EU Allowances Gold Standard for the Global Goals/ VER Verified Carbon Standard/VCUs UK Government Department of Energy and Climate Change Quality Assurance Scheme for Carbon Offsets	Approved standards/carbon credits: American Carbon Registry/ERTs Australian Emissions Reduction Fund/ ACCU Climate Action Reserve/CRTs Gold Standard for the Global Goals/ VER Kyoto Protocol's Clean Development Mechanism/CERs Kyoto Protocol's Joint Implementation/ ERUs Verified Carbon Standard/VCUs

(continued)

Table 5.2 (continued)

Steps	Australia NCOS	Carbon Footprint Standard	PAS 2060	The CarbonNeutral Protocol
Declaration	Publicly available reports should Detail progress against the emission management plan, the quantity and type of offsets purchased and the registry where they were retired. Where Australian offsets are used, full information must be publicly available and credits tracked on a public registry. Proponent must complete an agreement to use the National Carbon Offset Standard logo	Verification by Carbon Footprint Ltd. once certified, carbon footprint standard logo can be used in company's marketing materials under licence from Carbon Footprint Ltd. Licences are usually provided for a period of 12 months	Two forms of declaration: Commitment to carbon neutrality and achievement of carbon neutrality conformity assessment can be undertaken by an independent third-party certification, other party validation or self-validation	Self-validation not permitted; requirement and guidance on the use of the CarbonNeutral certification logo

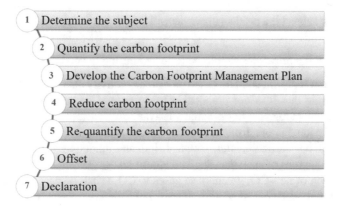

Fig. 5.1 Steps to demonstrate carbon neutral

When determining the subject for demonstrating carbon neutrality, PAS 2060 also requires the applicant to establish all characteristics (purposes, objectives or functionality) inherent to that subject.

Quantify the Carbon Footprint

The following international standards could be used to quantify carbon footprint of the subject:

- Organization/Entity: ISO 14064-1; GHG Protocol.
- Product/Service/Activity: PAS 2050; ISO 14067.
- Project: ISO 14064-2.

  According to PAS2060, the operational boundaries are defined as follows:

- All greenhouse gases shall be included and converted into CO_2e.
- 100% Scope 1 (direct) emissions included.
- 100% Scope 2 emissions included.
- Where estimates of GHG emissions are used in the quantification of the subject carbon footprint (particularly when associated with scope 3 emissions), these shall be determined in a manner that precludes underestimation.
- Any Scope 1, 2 or 3 emission estimated to be more than 1% of the total carbon footprint shall be taken into consideration unless evidence can be provided to demonstrate that such quantification would not be technically feasible or cost-effective.
- Emission sources estimated to constitute less than 1% may be excluded on that basis alone. All decisions to exclude shall be subject to the following conditions:
- The quantified carbon footprint shall cover at least 95% of the emissions from the subject.
- Where a single source contributes more than 50% of the total emissions, the 95% threshold applies to the remaining sources of emissions.
- Any exclusion and the reason for that exclusion shall be documented.
- When the subject is a product or service, all Scope 3 emissions should be included as the full life cycle from cradle to grave should be taken into consideration.

  The following principles should be followed during the quantification of CF:

- The carbon footprint shall be based on primary activity data unless the entity can demonstrate that it is not practicable to do so and an authoritative source of secondary data relevant to the subject is available.
- Use emission factors from national (Government) publications. Where such factors are not available, international or industry guidelines shall be used. In all cases the sources of such data shall be identified.
- Conversion of non-CO_2 greenhouse gases to CO_2e shall be based upon the 100-year global warming potential figures published by the IPCC or national (Government) publication.
- All carbon footprints shall be expressed as an absolute amount in tCO_2e. For products and services, these shall relate to a specified unit of product or instance of service (e.g. tCO_2e per kg of product).

Develop the Carbon Footprint Management Plan

According to PAS 2060, the company shall develop and document a carbon footprint management plan (CFMP) which shall include:

1. A statement of commitment to carbon neutrality for the defined subject.
2. A timescale for achieving carbon neutrality of the defined subject.
3. Targets for GHG reduction for the defined subject appropriate to the timescale for achieving carbon neutrality.
4. The planned means of achieving and maintaining GHG emissions reductions including assumptions made and any justification of the techniques and measures to be employed to reduce GHG emissions.
5. The offset strategy to be adopted including an estimate of the quantity of GHG emissions to be offset, the nature of the offsets and the likely number and type of credits.

CFMP is the deliverable from the carbon management strategic planning stage, as discussed in Chap. 4. CFMP shall be updated every year if the company wants to maintain its carbon neutrality status.

Reduce Carbon Footprint
The company shall implement the CFMP and monitor the progress against the plan over the time. Where the subject is a non-recurring event, the Plan shall identify ways of reducing GHG emissions to the maximum extent before the event takes place and include "post event review" to determine whether or not the expected minimization in emissions has been achieved.

Quantified GHG emissions reductions shall be expressed in absolute terms and shall relate to the application period selected and/or shall be expressed in emission intensity terms (e.g. per specified unit of product or instance of service).

It should be noted that the company shall reduce its emission as much as possible before seeking carbon offsetting. Hence according to PAS 2060, mandatory targets would be regarded as the minimum reduction target, and outsourcing could not regarded as a form of reduction, as all sources of emissions should be taken into consideration.

Re-quantify the Carbon Footprint
After reduction, carbon footprint should be quantified again at the end of each application period (i.e. reporting year). This is called residual carbon footprint, which is also the amount of GHG emissions that should be offset to achieve carbon neutral. The quantification methods are the same to what has been discussed above.

Offsetting
Company shall identify and document the standard and methodology used to achieve its carbon offset. The company shall prepare documentation substantiating the carbon offset including:

(a) Which GHG emissions have been offset.
(b) The actual amount of carbon offset.
(c) The type of offset and projects involved.
(d) Confirmation of the carbon offset scheme.
(e) The number and type of carbon offset credits used and the time period over which the credits have been generated.

(f) Information regarding the retirement/cancellation of carbon offset credits to prevent their use by others including a link to the registry where the offset has been retired.

Carbon offsets shall be verified by an independent third-party verifier. Other requirements include:

- Credits from carbon offset projects shall only be issued after the emission reduction associated with the offset project has taken place.
- Credits from carbon offset projects shall be retired within 12 months from the date of the declaration of achievement.
- Credits from carbon offset projects shall be supported by publically available project documentation on a registry which shall provide information about the offset project, quantification methodology and validation and verification procedures.
- Credits from carbon offset projects shall be stored and retired in an independent and credible registry.

Different offsetting standards and carbon credits can be adopted or approved by different carbon neutral standards, as summarized in Table 5.2.

Declaration
PAS 2060 does not make provision for a declaration of the achievement of carbon neutrality solely through offsetting other than the first application period. There are two forms of declaration:

- The declaration of *commitment* to carbon neutrality requires the entity to establish the carbon footprint and to document a carbon footprint management plan.
- The declaration of *achievement* of carbon neutrality requires the entity to have achieved reduction and to have offset remaining GHG emissions.

According to PAS 2060, validation can be undertaken by an independent third party, other party or through self-validation.

5.1.3 Certification of Carbon Neutral

In last section, overall process, steps and requirements based on different carbon neutral standards have been discussed. But how can a company have their company or products/service certified for carbon neutral? Different standards provide the different carbon neutral labels which are certified with their own standards, as shown in Fig. 5.2. For example, CarbonNeutral certifications can only be awarded by a CarbonNeutral certifier, following a certification process by which a client receives recognition that it has met the provisions of The CarbonNeutral Protocol for a specific subject.

PAS 2060 does not provide its own label, but many service providers adopt PAS 2060 standard to issue their own carbon neutral certification services with their own carbon neutral labels. A few examples can be seen from Fig. 5.3.

Based on PAS 2060:2014, there are three different ways to achieve carbon neutral certification, and the specified types of conformity assessment include:

– Self-validation.
– Other party validation.
– Independent third-party certification.

In the following section, examples of each above type will be discussed to give the readers an overview of the certification requirements and process.

Self-Certification
Marks and Spencer Group Plc. (Marks and Spencer 2014) issued the Qualifying Explanatory Statement (QES) to demonstrate Marks and Spencer Group Plc. (M&S) had achieved carbon neutrality from 1 January 2012 to 31 March 2014 and was committed to achieving carbon neutrality from 1 April 2014 to 31 March 2015 under the guidelines of PAS 2060:2014.

M&S conducted the self-certification in-house, and QES document did not use the QES checklist from PAS 2060:2014, but listing out its application periods, two declarations, carbon footprint assessment, carbon footprint management plan (CFMP), carbon offset strategy and detailed discussion on Scope 3 emissions, its exclusions and justifications.

Other Party Validation
Planet Labs manufactures and launches microsatellites, which provide regularly updated remote monitoring imagery around the world. As part of its commitment to environmental responsibility and climate protection, Planet Labs contracted SCS Global Services (SCS 2017) to certify its operations as carbon neutral in 2016 against the widely recognized PAS 2060 standard.

SCS issued QES for Planet Labs to declare its achievement of carbon neutrality at 31 December 2016 for the period commencing 1 January 2016 to 31 December in accordance with PAS 2060. An example of the checklist for QES supporting declaration of achievement to carbon neutrality is shown in Table 5.3.

Fig. 5.2 Carbon neutral labels certified with own standards; from left to right, Australia NCOS, Carbon Footprint Standard and The CarbonNeutral Protocol as described in Table 5.2

Fig. 5.3   Other examples of carbon neutral labels certified by PAS 2060 Standards

Third-Party Assurance

BP Target Neutral issued QES for Castrol Ltd. to demonstrate that it has achieved carbon neutrality for Castrol PCO engine oils and Castrol Engine Shampoo sold in Japan from 1 January 2018 to 31 January 2018 and was committed to maintain carbon neutrality from 1 February 2018 to 31 January 2019 (BP TN 2018). This QES is in accordance with PAS 2060 and was independently certified by a third-party Ernst & Young.

A more comprehensive QES checklist was also used for supporting declaration of commitment to carbon neutrality. For more details, readers could refer to PAS2060:2014 (BSI 2014).

5.2   Case Studies

5.2.1   Events

Many events have been claimed carbon neutral. Famous ones include 2008 Montreal Jazz Festival, US former president Obama's Inauguration in 2009, 2009 and 2010 Academic Awards, FIFA World Cup 2010 and 2014, Singapore's first carbon neutral event – the Responsible Business Forum on Sustainable Development (RBF) 2013, and Rio 2016 Olympic Games, etc. To conduct a carbon neutral event is a bit different from the standard process of measure-reduce-offset, as most events are non-occurring. The steps include:

- Pre-event assessment of carbon emissions.
- Planning and implementation of reduction measures.
- Post-event measurement of actual carbon emissions.
- Offsetting of residual emissions.

While mega sports events bring significant direct and indirect social and economic benefits to the host country, they also generate massive environmental impacts. Along the years, different mega events have tried to achieve carbon neutrality to show the event organizers' commitment in sustainability and

Table 5.3 Example of QES checklist requirements

☐ Define standard and methodology use to determine its GHG emissions reduction.
☐ Confirm that the methodology used was applied in accordance with its provisions and the principles set out in PAS 2060 were met.
☐ Provide justification for the selection of the methodologies chosen to quantify reductions in the carbon footprint, including all assumptions and calculations made and any assessments of uncertainty.
☐ Describe the means by which reductions have been achieved and any applicable assumptions or justifications.
☐ Ensure that there has been no change to the definition of the subject.
☐ Describe the actual reductions achieved in absolute and intensity terms and as a percentage of the original carbon footprint .
☐ State the baseline/qualification date.
☐ Record the percentage economic growth rate for the given application period used as a threshold for recognizing reductions in intensity terms.
☐ Provide an explanation for circumstances where a GHG reduction in intensity terms is accompanied by an increase in absolute terms for the determined subject.
☐ Select and document the standard and methodology used to achieve carbon offset.
Confirm that:
☐ (a) Offsets generated or allowance credits surrendered represent genuine, additional GHG emission reductions elsewhere,
☐ (b) Projects involved in delivering offsets meet the criteria of additionality, permanence, leakage and double counting,
☐ (c) Carbon offsets are verified by an independent third party verifier,
☐ (d) Credits from carbon offset projects are only issued after the emission reduction has taken place,
☐ (e) Credits from carbon offset projects are retired within 12 months from the date of the declaration of achievement,
☐ (f) Provision for event-related option of 36 months to be added here,
☐ (g) Credits from carbon offset projects are supported by publically available project documentation on a registry which shall provide information about the offset project, quantification methodology and validation and verification procedures,
☐ (h) Credits from carbon offset projects are stored and retired in an independent and credible registry.
Document the quantity of GHG emissions credits and the type and nature of credits actually purchased including the number and type of credits used and the time period over which credits were generated including:
☐ (a) Which GHG emissions have been offset,
☐ (b) the actual amount of carbon offset,
☐ (c) the type of credits and projects involved,
☐ (d) the number and type of carbon credits used and the time period over which the credits have been generated,
☐ (e) for events, a rationale to support any retirement of credits in excess of 12 months including details of any legacy emission savings taken into account,
☐ (f) Information regarding the retirement/cancellation of carbon credits to prevent their use by others including a link to the registry or equivalent publicly available record, where the credit has been retired,

(continued)

Table 5.3 (continued)

☐ Specify the type of conformity assessment: (a) independent third-party certification; (b) other party validation; (c) self-validation.
☐ Include statements of validation where declarations of achievement of carbon neutrality are validated by a third-party certifier or second-party organizations.
☐ Date the QES and have it signed by the senior representative of the entity concerned.
☐ Make QES publicly available and provide a reference to any freely accessible information upon which substantiation depends (e.g. via websites).

environmental stewardship. A comparison of four past events is summarized in Table 5.4.

Firstly, all these mega events conducted a pre-event assessment of carbon emissions. London Olympics defined it as "a reference footprint", which is a baseline assessment of what the games footprint would have been before efforts to reduce it (London 2012 2010). Performance is then assessed against this reference footprint using projected data which is adjusted to incorporate carbon reduction achievements and commitments.

As the standardized method for measuring and reporting CF for mega events was not available, London 2012 developed a well-documented specific methodology for Olympic and Paralympic Games 2012, which was also adopted in Rio 2016. For instance, for system boundaries, London 2012 combined both equity share and control approaches (approaches to define organizational boundaries for carbon measurement have been discussed in Chap. 2), to capture the emission both under its control and from those closely linked to its financial spend. The hybrid approach classifies operations as either "owned", "shared" or "associated" (OSA) according to the extent of the financial contribution from London 2012 Organising Committee (LOCOG) and the Olympic Delivery Authority (ODA). For London 2012, owned emissions from the ODA and LOCOG made up two-thirds of the overall, and the majority of owned emissions relate to the construction of venues and the Games-time operations.

2014 FIFA World Cup carbon footprint used operational control approach to define the organizational boundaries. It included preparation events, FIFA Confederations Cup (FCC) staging events and FIFA World Cup (FWC) staging events. It did not include the construction of the 12 stadiums used in the tournament – it only included the temporary constructions. FIFA 2018 used the same approach as FIFA 2014, but compared to the organizational boundaries set in 2014, the category "Team Workshop" has been added to the preparation period, and the International FIFA Fan Fests were excluded from the FWC period, since no International FIFA Fan Fests were planned to take place in 2018. For both FIFAs, transport was identified as the major contributor to total emissions: around 74% from international travel to Russia and travel between host cities for FIFA 2018 and 84% from transport at FIFA 2014.

Table 5.4 Comparison of carbon management and carbon neutrality for different mega sports events

	London 2012	FIFA 2014	Rio 2016	FIFA 2018
System boundaries and estimated CF	3.4MtCO$_2$e Venues (50%). Spectators (20%). Transport infrastructure (17%). Operations (13%)	2,723,756 tCO$_2$e Transportation (83.7%). Accommodation (5.7%). Venue (9.6%). Cross-phase activities (0.9%)	4.5MtCO$_2$e Operations (10%). Venue construction (16%). Infrastructure construction (19%). Spectator (55%).	2,167,118 tCO$_2$e Scope 1 (stationary combustion 0.4%). Scope 2 (electricity 1%). Scope 3 (temporary facilities 4.2%, F & B 4.9%, accommodation 11.7%, travel 73.8%).
Reduction efforts	More than 0.4MtCO$_2$e savings identified: Site energy strategy; venues re-design; materials specification; materials substitution; materials reuse/recycling; freighting by rail/water; fleet selection; technology selection; green procurement code	No strategy in reduction. Mainly focused on offsetting	Avoid emissions through careful planning and efficient processes. Reducing embodied carbon in materials through smart design and sustainable purchasing. Substituting fossil fuels for renewable and alternative fuels.	12 stadiums got BREEAM certified

(continued)

Table 5.4 (continued)

	London 2012	FIFA 2014	Rio 2016	FIFA 2018
Carbon offsetting	London 2012 dropped carbon offsetting	331,000 tCO$_2$e (including emissions from ticketholders who signed up to the climate protection campaign). Credits from four projects in Brazil	Engaged Dow as a climate partner to implement projects in Brazil and the rest of Latin America to achieve more than 2.2MtCO$_2$e during a 10 years' period. The Rio de Janeiro state government is responsible for the offset of 1.6 million CO$_2$e-related construction and infrastructure.	Launched a campaign to get ticket holders to offset the carbon emissions resulting from their attendance of the tournament
Post-event measurement	Actual emission reported as 3.3MtCO$_2$e	Nil	Nil	Nil

It should be noted that International Olympic Committee (IOC) released Carbon Footprint Methodology for Olympic Games in December 2018, which standardizes the method in CF measurement, management and reporting for mega sports event (ICO 2018).

Secondly, to reduce the impact, both Olympic Games London 2012 and Rio 2016 had developed a comprehensive sustainability plan to reduce and manage its carbon footprint. FIFA events did not have a solid reduction plan, except for the 12 stadiums that were certified green for FIFA 2018.

Thirdly, to offset the impact, London 2012 dropped the formal offsetting strategy as a carbon reduction solution. Therefore, London 2012 claimed to be the "greenest" event, but is not carbon neutral. Rio 2016 had a very complex compensation plan that it engaged Dow as a climate partner to develop and implement different carbon offsetting projects in Brazil and other Latin American countries. It also engaged the local government to offset part of the carbon emissions.

Although FIFA claimed carbon offsetting as the main reduction strategy and declared carbon neutrality for its events, neither FIFA 2014 nor FIFA 2018 could be certified as carbon neutral, in accordance with PAS 2060 or other standards. For example, FIFA 2018 launched its climate action campaign, encouraging successful ticket applicants to offset the carbon emissions resulting from their travel to the tournament for free. For each ticket holder signing up, FIFA offset 2.9 tCO$_2$e, which was the average emission per ticket holder travelling from abroad. However, the scheme was limited to only 100,000 tCO$_2$e – that is, it would only account for the emissions of 34,500 fans – out of three million viewers. The organization also pledged to offset all emissions considered "unavoidable" and operationally controlled, which is an amount of 243,000 tCO_2e, or only 11.2% of total emissions.

Finally, real impact assessment after the event. According to IOC's new methodology (ICO 2018), the results of the actual carbon footprint should be published as part of the post-Games Sustainability Report, normally within 6 months following the end of the Paralympic Games.

According to various post-event assessments, the London 2012 Games were regarded to be the greenest games ever, succeeding in reducing the carbon footprint of the event and cutting energy consumption by 20%. Although it fell short in other aspects, such as not meeting renewable energy targets, it was established that the carbon dioxide released during the games was 28% less than the projected amount. Even though the carbon footprint from transport was less than the projected levels, the carbon emission from spectators was estimated to be 913,000 Mt., which exceeded the expected emissions by approximately 36% (Environmental Leader 2012), caused by the higher number of spectators, athletes and the Olympics and Paralympics workforces. Nevertheless, the overall amount of emissions was at 3.3 million tCO_2e, which was lower than the estimate of 3.4 million tCO_2e (London 2012 2010).

The author could not find the post-event assessment for the actual carbon footprint for the other three events discussed here. When South Pole was asked if they would do an accounting of actual 2018 FIFA World Cup Russia carbon emissions to see how real emissions compare to the estimated ones, South Pole, who delivered the CF estimation for FIFA 2018, hoped they could be engaged by FIFA and the organizers of the 2022 World Cup in Qatar, where they could get the chance to report on the actual GHG emissions, and not just estimate them (see blog. southpole.com).

In summary, for small events to be claimed as carbon neutral, it is recommended to follow PAS 2060 and/or other international standards and also to follow the steps of pre-event assessment of CF, on-site implementation of reduction measures, post-event assessment of actual CF and offsetting the CF. For the mega events, it is recommended to stick to PAS 2060 and IOC's new guidelines on Carbon Footprint Methodology for Olympic Games.

5.2.2   Hotels

The Arthur Hotels group in Copenhagen was the first carbon neutral hotel group in the world (as confirmed by International Hotel & Restaurant Association, IHRA). Arthur Hotels group has developed and followed their 5-point climate action plan since 2008:

1. CO_2 neutralization now and in the future.
2. Create energy savings.

3. Involve guests.
4. Establish a CO_2 neutral hotel network.
5. Collaborate with climate networks/alliances including climate-friendly suppliers.

The most successful reduction measure by Arthur Hotels group was the reduction of linen consumption by 22%, which resulted in less laundry detergent used and reduced carbon emission from reduced energy consumption of the washing machines and reduced transport of linen to and from the hotel.

Hotel Speicher am Ziegelsee became the first carbon neutral hotel in Mecklenburg-Western Pomerania in 2010 and has been climate-positive since 2017. Guests could stay carbon neutral by the hotel's effort on 100% green electricity and 100% bio-district heating, promoting e-mobility and heat recovery in the entire irrigation technology. The measures helped to save 72.1% carbon emission per guest per night and to consume 59.1% less energy in 2017 than hotels in the same star category during the same period. Avoiding long-distance delivery, the hotel sources predominantly regional products. Residual emissions are compensated by a reforestation project in Panama as part of the initiative "Klima-Hotels Deutschland" (https://www.green-pearls.com/newsroom/climate-neutral-hotels/).

The Bucuti & Tara Beach Resort in Aruba has become the first in the Caribbean to achieve total carbon neutrality in 2018. It followed the CarbonNeutral Protocol and was certified as CarbonNeutral. It implemented much broader reduction measures including daily transportation for staff and the importation of food and goods, as well as any carbon spent for conferences and business travel. The resort is home to the largest solar panel installation the government of Aruba will allow, and the small amounts of offsets they need to purchase come from a local wind farm. They also make efforts in waste reduction, especially engaging the guests in diverting single-use plastic bottled and food waste from landfilling.

In Hong Kong, iclub Wanchai Hotel (formerly named "Regal iclub Hotel") is a hotel brand under Regal Hotels International. During my time at Carbon Care Asia, we have assisted the iclub Wanchai Hotel to become Hong Kong's first carbon neutral hotel in 2010. We established a comprehensive carbon management system through "measuring, reducing and offsetting" since the hotel's opening, developed carbon reduction and offsetting strategies and provided communications support to stakeholder messages on "carbon neutrality".

For the hotel sector, in the past, hotels often did not go beyond their own operations of the hotels when talking about carbon neutrality. But a trend has been seen recently that hotels are looking into the carbon emissions from Scope 3, especially those related to goods and services purchased by hotels, waste management and transportation of the guests, staff and purchased goods, all of which are required by PAS 2060. Hotels are also providing a good platform to engage their guests and raise their awareness on climate and other environmental impacts. The hotel's carbon footprint and the footprint per guest and stay should be kept as low as possible through initiatives and latest technology. The detailed case study on hotel carbon management will be discussed in Chap. 6.

5.2.3   Corporate

Business for Social Responsibility listed out the carbon neutral companies and projects in its report in 2007 (BSR 2007). Table 5.5 summarizes a few selected companies. I will take HSBC as an example in this section.

In 2004 HSBC made public its intention to go carbon neutral by January 2006 and achieved this objective, becoming the world's first carbon neutral bank in October 2005, 3 months ahead of schedule. HSBC's Carbon Management Plan comprised four steps to achieve carbon neutrality (HSBC 2008):

- Measuring carbon footprint.
- Reducing emissions from energy consumption and business travel.
- Buying renewable energy where possible.
- Offsetting any remaining carbon emissions.

In 2006, measured CF for HSBC was 813,000 tCO_2e, among which 69% came from its purchased electricity, 8% from fuel used for its operations and 23% from business travels. In 2005, HSBC set 3-year targets to reduce its energy consumption by 7% and carbon emissions by 5%. A number of initiatives have been implemented

Table 5.5 List of some carbon neutral companies

Target Year	Company	Revenues ($ millions)	Industry	Status
2000	Shaklee	34	Personal & home care	Achieved
2005	HSBC	115,361	Banking/finance	Achieved
2006	Barclays UK	47,942	Banking	Achieved
	British sky broadcasting	7,534	Media	Achieved
	World Bank	4,783	International NGO	Achieved
	Avis Europe	1,512	Auto rental	Achieved
	BSI	467	Standards	Achieved
	Simmons & Simmons	449	Legal	Achieved
	Silverjet	N/A	Airlines	Achieved
2007	Bradford and Bingley	3,853	Financial services	Achieved
	Green Mountain power	241	Electricity generation	Achieved
	Salesforce.com	497	Professional services	Commitment
	Yahoo! Inc.	6,426	IT	Commitment
2008	KPMG (Australia)	462	Accounting services	Commitment
2010	News Corp	25,327	Media	Commitment
	ST microelectronics	9,854	Semiconductors	Commitment
	Timberland	1,567	Apparel/footwear	Commitment
2012	Marks & Spencer	13,561	Retail	Commitment
	Nike	14,955	Apparel/footwear	Commitment
2020	Interface, Inc.	1,076	Commercial interiors	Commitment

to achieve these reduction targets. It included building a prototype zero carbon branch in New York, installing solar energies for buildings in the UK and France, installing videoconferencing technology in a number of offices to reduce the need for employee business travel and rolling out a few energy efficiency initiatives. With step 3, HSBC buys green energy in many regions around the world. For instance, in 2006, HSBC bought clean energy equivalent to a third of its electricity consumption in the USA through the purchase of renewable energy credits. And lastly, HSBC bought carbon dioxide offsets from credible renewable energy projects which have been assessed and verified independently.

In July 2007, HSBC committed to spend US$90 million over the next 5 years to continue to reduce its carbon footprint. This Global Environmental Efficiency Programme would help the Group achieve its environmental reduction targets by trialling environmental innovation and sharing best practice through the installation of renewable energy technologies and other initiatives (HSBC 2008).

Furthermore in 2017 HSBC launched the HSBC Climate Partnership programme to commit US$100 million to work with The Climate Group, Earthwatch, the Smithsonian Tropical Research Institute and the World Wide Fund for Nature to combat climate change by inspiring individuals, business and governments worldwide. In this partnership programme, HSBC employees were aimed to be engaged through the Climate Champion programme under which trainings on sustainability and climate change were provided to equip them to initiate changes in their own business areas (Hopwood et al. 2010).

However, in 2011, HSBC announced it would no longer be carbon neutral, because the carbon market didn't work out as expected when it made the commitment in 2005 (Bowen 2014). In 2013, in a media briefing published by World Development Movement, it blamed that HSBC's investment banking arm was the UK's biggest underwriter of fossil fuel bonds and shares, and between 2010 and 2012, HSBC helped fossil fuel companies raise just under £75 billion (World Development Movement 2013). It also blamed that HSBC's carbon neutrality declaration did not consider any carbon emissions from bankrolling fossil fuel firms to form part of its carbon footprint. HSBC's previous supposed carbon neutrality only referred to its annual spend of £ten million a year on carbon offset schemes and other projects, while £ten million is less than 0.0007% of HSBC's overall assets billion (World Development Movement 2013).

On 6 November 2017, HSBC had a news release that it pledged to provide $100 billion in sustainable financing and investment by 2025, as part of its new commitments to tackle climate change and support sustainable growth in the communities it serves. HSBC pledged to intensify its support for clean energy and lower-carbon technologies, as well as projects that support the implementation of the United Nation's SDGs, which included sourcing 100% of its electricity from renewable sources by 2030, with an interim target of 90 per cent by 2025 and discontinuing financing of new coal-fired power plants in developed markets and of thermal coal mines worldwide.

5.2.4   Products

O'right Carbon Neutral Shampoo
Based in Taiwan, O'right is the maker of the world's first carbon neutral shampoo. Since 2009, O'right has asked SGS Taiwan to conduct the product CF of three signature products – 400 ml Green Tea Shampoo, 1000 ml Green Tea Shampoo and Organic White Tea Hair Treatment Set, following PAS 2050:2008. In 2010, O'right participated in Taiwan's newly launched Carbon Footprint Label system and got the world's first carbon neutral shampoo certified by BSI.

Based on O'right's carbon neutral declaration report (O'right 2011), the CF of a bottle of 400 ml Green Tea Shampoo was around 790 gCO_2e from its life cycle, among which raw material extraction contributed 19%, manufacturing 3%, logistic and retail 2%, consumer 50% and waste disposal 26%. CF was reduced by 2.08% due to the change of the raw materials and new design of the bottle and the residual of 8 tCO_2e for its 10,000 bottles of shampoo produced from 1 January 2011 to 30 June 2011 were offset by purchased VCUs.

O'right moved forward beyond 2011. In 2016, it developed the groundbreaking 100% renewable plastic shampoo bottle made from recycled food and cosmetic plastic containers. In 2017, all 1000 ml conditioner bottles were upgraded to 100% renewable plastic to achieve a carbon reduction of 75%. In 2018, O'right brought sustainability to the next level by successfully joining RE100 along with Facebook and Google and vowed to become the first company in Taiwan to make a commitment to 100% renewable electricity by 2025. The same year, O'right was validated by SGS for carbon neutrality as a corporate organization and for nine of their products and introduced the world's first renewable plastic pump.

Interface Carbon Neutral Floor
Interface, a US flooring giant, has continuously been recognized as the corporate sustainability leader in GlobeScan and SustainAbility's annual Sustainability Leaders Survey since 1997. Interface is well-known for its commitment on its "Mission Zero" pledges – zero waste, greenhouse gas emissions and net water use – by 2020, and in 2016 it launched a new Climate Take Back mission, which sets a series of innovation goals to inspire both Interface and other businesses to further reduce their environmental footprint and even help reverse climate change.

In 2018, every flooring product offered by Interface – whether carpet, LVT or rubber – is 100% carbon neutral. Carbon emission from every stage of its products' life cycle has been analysed, and Interface reduced emissions of manufacturing by 96% and reduced the carbon footprint of its products by over 60%, which remains the lowest in the industry. Interface has purchased more than 3.9 million tonnes of carbon credits since 2002 to offset its emission and to support various RE and community projects, as listed below:

- Madagascar, Thailand and China – new renewable energy projects involving solar, hydro- and wind power to reduce the amount of carbon emissions entering the atmosphere.

- Guatemala and Kenya – community fuel switching and water purification projects tackling carbon emissions alongside human health and community empowerment.
- USA, Cambodia and Zimbabwe – reforestation projects to keep carbon stored and sequestered in the soil and plants.

FIJI Water

In 2007, FIJI Water announced a plan to be carbon negative – that is, to trap more GHGs than it released in the process of making, shipping and selling its product, bottled FIJI Water (Deutsch 2007). The overall goal was to reduce emissions from its own operations and offset more emissions than its residual emission so that FIJI bottled water was a 20% carbon-negative product. The company planned to achieve this through a number of methods, including the installation of a windmill on its plant in Fiji, the use of more ships and fewer trucks to transport its products, reducing the amount of plastic and paper used in packaging and increasing the amount of alternative fuels used in its trucks and at its plant. In addition to these internal changes, the company would also purchase carbon offsets to reduce emissions and announced a partnership with Conservation International to permanently protect the 50,000-acre Sovi Basin, the largest lowland rainforest remaining in Fiji (GreenBiz 2007).

FIJI Water was the first private company and the first bottled water company to disclose its CF on the Carbon Disclosure Project (now named CDP) (Gino et al. 2013). ICF International, a global leader in analysing emissions inventories and providing advice on climate strategy, independently reviewed and verified FIJI Water's carbon footprint annually. According to an interview with FJI Sustainability Manager Barbara Chung by Inhabitat, purchasing 1-litre bottle of FIJI Water resulted in removal of about 115 g CO_2e, and FIJI Water in 2008 helped remove more than 20,000 tCO_2e from the atmosphere, which is equivalent to planting over 500,000 trees (Inhabitat 2008).

In 2011, a Southern California woman sued the FIJI Water Company in a class action complaint that alleged the company's carbon-negative claim was deceptive and misleading. The law firm Newport Trial Group filed the suit on December 20 in the US District Court in California. It stated that FIJI Water used forward offset credits, which relied on future carbon reduction that may or may not take place, and clients paid more due to its carbon-negative claim.

Forward crediting occurs when one purchases the carbon credits that will be generated in the future. It has been mentioned in Section 3.4.2 when discussing the controversial issues of afforestation projects. For instance, under Gold Standard (refer to Table 3.4), tree planting projects are excluded. So it really depends on which offset standards and which types of the projects the company will adopt and how the company measures and makes the marketing claim of its products' carbon neutrality status.

5.2.5   City

Masdar City is a sustainable city project hosted by Abu Dhabi, the capital of the United Arab Emirates. In 2008, it announced to be the world's first zero carbon zero waste city. The planned development was a dense, car-free city to be constructed in an energy-efficient two-stage phase. A 10 MW photovoltaic solar power plant, the largest in the Middle East, would be constructed to power the city (Teagarden 2017). However, by 2016, only about 5% of Masdar's original plan 6 km2 footprint with 50,000 inhabitants have been finished. Project planners have extended the completion date from 2016 to 2030 but gave up on the zero carbon zero waste goal, as they thought it would not be economically and technologically viable (Teagarden 2017).

A few other projects have also been piloted or targeted itself for carbon neutral. Readers could have a read of unsuccessful projects such as Dongtan (C40 2011; McGirk 2015) and successful trial of energy-positive 4000-inhabitant Danish island of Samsø, where, during the past decade, more energy has been produced from wind and biomass than it consumed (Lewis 2017).

How about commitment by cities after Paris climate conference, especially for big cities or even mega cities?

The Carbon Neutral Cities Alliance (CNCA) is a collaboration of leading global cities working to cut greenhouse gas emissions by 80–100% by 2050 or sooner – the most aggressive GHG reduction targets undertaken anywhere by any city. CNCA develops approaches, analysis, and tools to support carbon neutrality planning and implementation and standardizes measurement and verification methodologies in order to track progress. It developed Framework for Long-Term Deep Carbon Reduction Planning and its Game Changer Project to support city policies and initiatives that have the highest potential for rapid, deep GHG emissions reductions in urban transportation, energy use, and waste systems. The project also shares best practices for achieving "transformative" deep carbon reduction strategies.

Based on the information on the CNCA's website, the author summarizes the carbon neutral targets and programmes of some CNCA member cities, as shown in Table 5.6. More information could be found from CNCA's website: https:// carbonneutralcities.org/. The author summarizes the leader cities and their carbon targets as follows:

Past leaders:

– The city of Melbourne became a certified carbon neutral organization for the first time in 2011–2012.
– The City of Sydney's operations became carbon neutral in 2007, with the City being the first government in Australia certified as such in 2011.

Current leaders who set the carbon neutral targets:

– Melbourne (2020).
– Adelaide (2025).
– Copenhagen (2025) – aims to be the first carbon neutral capital by 2025.
– Stockholm (2040).

Cities who set carbon neutral target by 2050:

– Berlin, San Francisco, Seattle, Sydney and Washington DC.

Cities who set emission reduction target of 80% by 2050:

– 1990 as baseline: Portland and Toronto
– 2005 as baseline: Boulder, New York, Rio De Janeiro and Yokohama
– 2006 as baseline: Minneapolis
– 2007 as baseline: Vancouver

Renewable Energy Commitments:

– Oslo already achieved 100% RE.
– 100% RE by 2025: Copenhagen.
– 100% RE by 2030: Boulder.
– 100% RE by 2050: Portland and Vancouver.

5.3   Building Climate Resilience

In Chap. 1, when we discussed climate change consequences, the climate risks and impacts and the concepts of mitigation and adaptation have also been introduced (see Sect. 1.4). But the concept of carbon management discussed so far is mainly based on the mitigation, that is, to understand where and how emission come from (i.e. carbon measurement) and to minimize it (i.e. carbon reduction and offsetting). IPCC in its AR5 projected the different climate changes; an adopted figure is shown in Fig. 5.4 (IPCC 2014a).

The Representative Concentration Pathways (RCPs) describe four different pathways of GHG concentrations for the twenty-first century:

– RCP2.6: a stringent mitigation scenario.
– RCP4.5 and RCP6.0: two intermediate scenarios.
– RCP8.5: a very high emissions scenario.

Scenarios without additional efforts to constrain emissions ("baseline scenarios") lead to pathways ranging between RCP6.0 and RCP8.5. RCP2.6 represents a scenario to keep global warming likely below 2 °C above pre-industrial temperatures.

Now the question is how business can act to the projected climate risks? IPCC defines resilience as "the ability of a system and its component parts to anticipate, absorb, accommodate, or recover from the effects of a hazardous event in a timely and efficient manner, including through ensuring the preservation, restoration, or improvement of its essential basic structures and functions" (IPCC 2012). In a BSR's 2018 report, a climate resilient business is therefore defined "be able to

Table 5.6 Selected CNCA member cities' carbon neutral targets and programmes

City	Target	Policies and strategies	Programmes
Adelaide	Carbon neutral for community by 2025 and for its operations by 2020	Carbon Neutral Strategy 2015–2025	Carbon Neutral Adelaide Action Plan 2016–2021 outlines a way forward for mobilizing efforts to achieve carbon neutrality
Berlin	Climate neutral by 2050; reduce its emissions by 85% by 2050 vs.1990 with a reduction of at least 40% by 2020 and at least 60% by 2030	Berlin Energy Turnaround Act as amended in 2017; Berlin Energy and Climate Protection Programme 2030, with approval by the Berlin Senate and Parliament	Ending coal-based generation: Lignite-based power has been phased out in 2017, and black coal-based power will be by 2030; move away from combustion engine technology, and reduce emissions from motorized private transport; public administration will work carbon neutral by 2030
Boulder	Reduce 80% of community-wide emissions from 2005 by 2050; reduce emissions from city operations 80% below 2008 by 2030; and achieve 100% RE community-wide by 2030	Climate commitment document, the adopted goals also include progress indicators and targets for local RE generation, energy efficiency, electric vehicle adoption and waste and water reductions for key milestone years	Building performance ordinance for efficiency in commercial and industrial buildings, updating energy codes, SmartRegs rental housing efficiency requirements, one-on-one business and residential advising and reducing landfill emissions through the implementation of the universal zero waste ordinance
Copenhagen	Be the first carbon neutral capital in 2025	CPH 2025 Climate Plan adopted in 2012	Four pillars: Energy consumption; energy production; mobility; city administration initiatives
Melbourne	Set the municipal target of zero net emissions by 2020		Innovative 1200 Buildings, Smart Blocks, City Switch and Solar Programs provide information and solutions and address barriers to reducing emissions
New York	80% reduction by 2050	Mandatory retrofits to city buildings and expanding low-carbon transportation options	Investing over $20 billion to adapt to climate change risks. Divesting the City's pension funds from fossil fuels and suing the five investor-owned fossil fuel companies
Washington, DC	Become carbon neutral by 2050	Climate ready DC plan; clean energy DC	Cut emissions by 50% by 2032 by cutting energy use and increasing the use of renewable energy

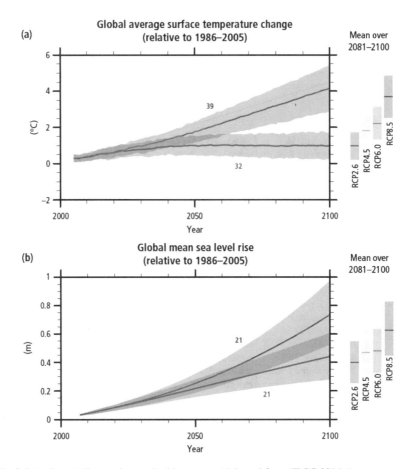

Fig. 5.4 Projected climate changes in this century. (Adopted from (IPCC 2014a))

anticipate, absorb, accommodate, and rapidly recover from climate events in its own operations and throughout its value chain" (Cameron et al. 2018). Marsh & McLennan Companies' Global Risk Center further extends that climate resilience is "the capacity not only to survive, but also to adapt and succeed in the face of climate change and its direct and indirect impacts, including changes in regulation and policy" (Nottingham and Yeo 2017). In this section, I will discuss how company could build its climate resilience.

Before taking actions, corporate management and the board first must develop a robust view of how climate change impacts – directly and indirectly – affect the business performance, company operations and financial implications. So to build climate resilience, it starts with the assessment of the climate risks and vulnerabilities, the insights from which could help support organizations' decision-making process concerning capital allocations, operation management and risk mitigation.

5.3.1 Overall Approach

Companies should always leverage their existing enterprise risk management (ERM) system and process to assess the climate risks and develop adaptive measures and add them into the risk registry. ISO 31000:2018, Risk management – Guidelines provides principles, framework and a process for managing risk, which include setting the context, risk assessment and risk treatment.

Companies could also use different methods during this risk management process, especially related to climate risk and vulnerability assessment (USAID 2018):

- Desktop reviews to synthesize information from existing resources.
- Stakeholder consultations and workshops to obtain input through interviews, roundtables or workshops on the impacts of climate and other factors in determining vulnerabilities.
- Additional analyses to determine and characterize climate hazards, vulnerabilities or risks in greater detail. Examples of additional analyses are hazard, vulnerability or risk mapping, impact modelling, institutional assessment and economic impact analysis.

By combining these methods, the overall approach developed by the author based on her consultancy jobs is illustrated in Fig. 5.5. There are four stages as listed below:

- Setting the context

In this stage, the consultant and client shall agree the project approach, scope and assessment framework; literature review on climate impacts for specific sector and climate scenarios for specific geographic location also shall be conducted at this stage.
- Risk assessment

This includes climate risk assessment and vulnerability assessment. The consultant shall help the client to prepare climate risks list and conduct stakeholder engagement to confirm the vulnerabilities. Further risk assessment, prioritization and evaluation shall also be conducted at this stage.
- Risk treatment

The next step is to develop adaptation plan and integrate it with the mitigation measures to building climate resilience. Monitoring and reporting systems shall also be developed in this stage. It is highly recommended that climate risk could be incorporated into the company's risk registry and be managed by using the same ERM system.

• Knowledge development

Lastly, consultant shall provide necessary trainings for the client to build enough knowledge to continue the exercise by themselves. The consultant and client could also conduct further research on the new and innovative technologies to better adapt to the dynamic climate risks.

Case: New World Development's Climate Resilience Approach

Based on New World Development (NWD) Company Limited's 2018 Sustainability Report titled "Our Vision Your New World", below summarizes how NWD builds its climate resilience.

On Governance:

– Climate risks have been incorporated into Group's Risk Management and Internal Control Assessment Checklist.
– The Board oversees climate risks, which form part of the Risk Management Policy.
– The Audit Committee decides the Group's overall risk level, which includes climate-related risks, and ensures the effectiveness of its risk management system.

On Approach:

The below figure illustrates how NWD embeds climate risks into its sustainability reporting exercises, such as sustainability materiality test, and the ERM and risk control checklist.

List of Climate Risks Identified and Assessed	Sustainability Materiality Test	ERM and Risk Control Checklist	Capacity Building	Biannual Review and Continuous Improvement
Internal sustainability team or external consultant	identify stakeholders' priorities on key sustainability topics, business opportunities and risks (internal+ external)	Engages all NWD departments and business units to identify, assess and address these risks (internal)	Group-wide ESG training to raise internal capacity in managing relevant risks and performance	Audit Committee conducts a biannual review of ESG risks identified by all internal stakeholders via the Checklist

5.3.2 Climate Risk Assessment

List Out Climate Hazard
Climate hazard: the potential occurrence of a climate-related physical event or trend or their physical impact that may cause loss of life, injury or other health impacts, as well as damage and loss to property, infrastructure, livelihoods, service provision, ecosystems and environmental resources.

C40 classifies climate hazards into five categories, as shown in Table 5.7, which is a good starting point of climate risk assessment, providing the company assessed is located in urban areas. It also emphasizes that by the end of the twenty-first century, the key trends in climate conditions that may affect climate hazards are likely to include (C40 2015):

- Rising average temperatures.
- Increasing frequency and intensity of extreme heat.

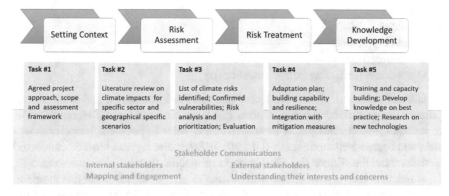

Fig. 5.5 Overall approach of building climate resilience

Table 5.7 List of city climate hazard (C40 2015)

Category	Hazard	City climate hazard
Meteorological	Precipitation, wind, lightning, fog, extreme cold and extreme hot	Storm, heavy snow, severe wind, tornado, typhoon, electrical storm, fog, extreme winter, heat wave, extreme hot weather
Climatological	Water scarcity, wild fire	Drought, forest fire, land fire
Hydrological	Flood, wave action, chemical change	Surface flood, river flood, coastal flood, groundwater flood, storm surge, salt water intrusion, ocean acidification
Geophysical	Mass movement	Landslide, avalanche, rockfall, subsidence
Biological	Insects and microorganisms	Water-borne disease, vector-borne disease, air-borne disease, insect infestation

- Precipitation variability and extreme precipitation events.
- Sea level rise.

It should be noted that hazard is geographically specified; for instance, in Hong Kong, the probability of heavy snow or rockfall is very low. But the climate change is dynamic, and extreme changes to the local environment may affect the frequency and severity of the climate hazard. Literature reviews shall be conducted to understand the projection of the local climate data, especially based on different RCP scenarios.

For business organizations, the impact areas could be defined as:

Assess Climate Impacts
Climate impacts: effects on natural and human systems of climate hazard. Impacts generally refer to effects on lives, livelihoods, health, ecosystems, economies, societies, cultures, services and infrastructure due to the interaction of climate changes or hazardous climate events occurring within a specific time period and the vulnerability of an exposed society or system.

- Facilities.
- Financial.
- Legal/compliance.
- Market/reputation.
- Operational (including human resources).

Climate risks and impacts are sector specific. For instance, for real estate sector, potential impact of the extreme weather like typhoon could include increased costs to repair or replace damaged or destroyed assets, value impairment, property downtime and business disruption, etc. (ULI 2019). Literature reviews shall be conducted to understand the climate risks and impacts in specific sector under assessment.

A risk heat map is a tool that could be used to present the results at this moment. It represents the evaluations of the likelihood or probability of a hazard and the severity of its impact. An example of risk heat map matrix is shown in Table 5.8.

Table 5.8 Example of risk heat map

Likelihood Severity	Very unlikely (<10%) ①	Unlikely (10-33%) ②	Moderate (33%-66%) ③	Likely (66%- 90%) ④	Very likely (>90%) ⑤
Extreme ⑤	5	10	15	20	25
High ④	4	8	12	16	20
Medium ③	3	6	9	12	15
Low ②	2	4	6	8	10
Negligible ①	1	2	3	4	5

Company could define its own calculation method for assessment. For instance, in Table 5.8, the likelihood is defined based on IPCC's probability definition. For an acute event, frequency is normally used to present likelihood.

To facilitate the calculation, scores could be defined for each level of severity or likelihood. For instance, in this 5 × 5 heat map, if 1 to 5 points are assigned to each level of the likelihood (L) and severity (S), the risk impact score can be calculated as L × S, as shown in Table 5.8. Higher score means identified risks with more critical impact.

Climate Vulnerability Assessment

The extent of the impact depends on the magnitude of climatic changes affecting a particular system (exposure), the characteristics of the system (sensitivity) and the ability of people and ecosystems to deal with the resulting effects (adaptive capacities of the system). These three factors determine the vulnerability of the system, as shown in Fig. 5.6.

Exposure – the presence of people, livelihoods, species or ecosystems; environmental functions, services and resources; infrastructure; or economic, social or cultural assets in places and settings that could be adversely affected.

Sensitivity – the degree to which a system is affected, either adversely or beneficially, by climate variability or change.

Adaptive capacity – the ability of a system to adjust to climate change (including climate variability and extremes) to moderate potential damages, to take advantage of opportunities or to cope with the consequences.

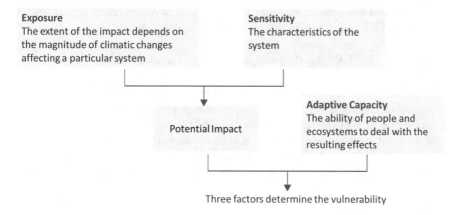

Fig. 5.6 Vulnerability determined by exposure, sensitivity and adaptive capacity

Climate hazard such as extreme weather events could have devastating effects on vulnerable property and critical infrastructure, with lasting impacts across companies and its communities. There are different methods for vulnerability assessment as either following top-down or bottom-up approaches. Top-down approaches start with an analysis of climate change and its impacts, while bottom-up approaches start with an analysis of the people affected by climate change (GIZ 2014).

By using top-down approach, after conducting geographic portfolio review of the climate hazard for the company, the next step is to map demographic and infrastructure vulnerabilities to natural hazards and thereby identify the aggregated weather exposure with respect to location, facility and asset.

Bottom-up approaches start with underlying development context of why people are sensitive and exposed to a given climate hazard like flood. Sensitivity to climatic change is generally high when societies depend on natural resources or ecosystems, e.g. agriculture and coastal zones and poor communities are especially vulnerable to climate change due to their limited access to resources, secure housing, proper infrastructure, insurance, technology and information (GIZ 2014).

Vulnerability (V = Exposure + Sensitivity – Adaptive Capacity) could then be assessed by using the defined weighting method based on analytic hierarchy process (Hammill et al. 2013).

Assess and Prioritize the Climate Risks
Risk: The potential for consequences where something of value is at stake and where the outcome is uncertain, recognizing the diversity of values. Risk is often represented as probability of occurrence of hazard or trends multiplied by the impacts if these events or trends occur. Climate risk refers to the risks of climate impacts.

Risk results from the interaction of vulnerability, exposure and hazard. Climate risk rating (CRR) can be assessed by the following equation:

$$CRR = [Likelihood \times Severity] \times [Vulnerability]$$

If Likelihood scored from 1 to 5, severity scored from 1 to 5 and vulnerability scored from 1 to 10, then the CRR could be assessed by Table 5.9.

A general idea and process to assess climate hazards, vulnerabilities and climate risk has been introduced in this section. It has to emphasize again that results from climate risk assessment should be incorporated into company's enterprise risk management (ERM) framework and system. It also should be noted that company

Table 5.9 Climate risk rating

CRR	Extreme	High	Medium	Low	Negligible
Score	>200	150–200	100–150	50–100	<50

could have its unique method for risk assessment such as setting specific scores or thresholds for consequences and likelihood, selecting the difference approach for vulnerabilities assessment or using different method in generating risk ratings.

5.3.3 Adaptive Measures

After identifying and prioritizing risks and opportunities, the next step is to develop adaptation plan to enable company to respond and against future risks, to reduce the negative climate impacts of climatic change and to prepare for opportunities in future. Companies can apply a variety of instruments in their risk-mitigation toolkit to enhance their physical, operational and financial resilience.

BSR proposed a framework to address six capitals to reduce the climate risks and build climate resilience (Cameron et al. 2018). It suggested that some climate impacts can be absorbed and accommodated if adaptive capacity is strengthened, using six so-called capital assets: human, financial, social, natural, physical and political capital.

- Human capital refers to the skills and knowledge in the workplace.
- Financial capital refers to the available financial resources and access to financial goods and services.
- Natural capital refers to the products and services provided by natural resources, biodiversity and ecosystems.
- Political capital refers to the access to decision-making to shape policy that enable resilience.
- Physical capital refers to infrastructure and equipment, including facilities, transport, logistics and communications.
- Social capital refers to collaborations, partnership and support.

These six capitals are considered the key building blocks of resilience. When companies evaluate the climate risks and understand the external scenarios such as the changing climate hazards and evolving political, social and regulatory environments, company would establish internal plans to strength its adaptive capacity, design the resilience strategy with investment in six capital assets and engage internal and external stakeholders to forge effective collaboration to drive implementations.

5.3.4 Interaction Between Adaptation and Mitigation

Based on the above discussion, company could identify means of increasing its climate resilience through direct physical risk mitigation, such as infrastructure reinforcement in coastal areas, or by implementing sustainable supply chains and operational processes. Company could also develop strategies to address the transition risks which include regulatory changes, customer demands shifts and the changing availability and price of resources due to a shift to a lower-carbon economy and using new, non-fossil-fuel sources of energy. By ensuring that physical and transition climate risks are incorporated into a company's risk register and management programs, company can identify optimal responses and opportunities to improve corporate performance and financial earnings.

The last question left is how about the interaction between mitigation measures, which aim to reduce GHG emissions, and adaptation measures, which aim to reduce climate risks? For example, carbon-reduction strategies are often deployed under considerations of resource-constraint risks or regulatory risks.

Funded by the Children's Investment Fund Foundation, C40 developed a tool called Adaptation and Mitigation Interaction Assessment (AMIA) to help cities to understand the relation between mitigation and adaptation (C40 2018). When users provide a list of mitigation and adaptation actions, AMIA tool is designed to assess the interaction between two in four areas:

- Synergies: the win-win situation, when actions reduce both carbon emissions and climate risks.
- Trade-offs: when actions have contrary effects on adaptation and mitigation, so when mitigation actions increase climate risk or adaptation actions increase emissions.
- Risks of mal-investment: when actions can be undone or rendered less effective by the effects of climate change if they are not sufficiently resilient.
- Piggybacking opportunities: when actions are coupled in their design or implementation and additional mitigation or adaptation actions are added at a small marginal cost.

Table 5.10 gives an example of the results obtained from AMIA tool. In this example, flooding is the selected climate hazard. Three adaptive measures selected as storm surge barriers, pumping stations and multifunctional flood defences and two mitigation measures are building energy operational improvements and EV charging infrastructure.

5.4 Total Carbon Management Model

To summarize this chapter, the total carbon management model is illustrated in Fig. 5.7. When Paris Agreement was adopted by around 195 nations in 2015, it showed that most of the world nations accepted IPCC's findings based on its AR5, and the Earth requested collective will and efforts in keeping temperature increase below 2 °C of the pre-industrial level.

It means two aspects to a business: on the one hand is to meet the reduction target assigned by the sector or set more ambitious target to contribute more reductions; on the other side is to build its own climate resilience to better anticipate and accommodate the climate risks and better respond and recover from it.

Table 5.10 An example of interactive assessment between adaptation and mitigations

Action Name	Description	Synergy	Potential trade-off	Mal-investment risk	Piggybacking opportunities
Adaptation #1 Multifunctional flood defences	This land can be used for other functions. For example, use the dikes for constructing park	Defence that integrates roadways and mobility solutions that can reduce emissions through eased traffic congestion	Water pumping associated with multifunctional flood defences may increase energy use and GHG emissions.		Multifunctional flood defences can have additional mitigation measures integrated into the design, such as low-emission transport solutions or tidal power
Adaptation #2 Pumping stations	Used to pump away water in case of floods or rainfall		Water pumping can increase energy use and associated GHG emissions		Distributed solar energy, particularly rooftop PV, or in certain cases pumped hydroelectric storage, which can provide mitigation benefits
Adaptation #3 Storm surge barriers	Storm surge barriers are large structures that can close off a waterway in case of storms or high water	Integrated pedestrian walkways, cycling and other mobility solutions can reduce GHG emissions	Construction activities and materials form storm surge barriers will have associated GHG emissions		Storm surge barriers can have tidal power turbines integrated into the barriers
Mitigation #1 Building energy efficiency – Building energy operational improvements	Implement improved operational requirements for buildings to reduce energy use	Reduce GHG emissions and also reduce peak demand and improve grid resiliency during extreme heat			Building operational improvements for energy efficiency present opportunities to integrate adaptation measures at the same time, such as water efficiency improvements

(continued)

Table 5.10   (continued)

Action Name	Description	Synergy	Potential trade-off	Mal-investment risk	Piggybacking opportunities
Mitigation #2 Electric vehicle charging infrastructure	Help incentivize the adoption of electric vehicles by the public	Increase fuel supply resiliency through diversification of transit fuel types. Reduce urban heat island effect via reduced criteria pollutant emissions		EV charging infrastructure vulnerable to damage during flood events	New EV charging infrastructure can incorporate adaptation measures such as water proofing. It can also present an opportunity to integrate shading and urban greening, the construction of adaptation infrastructure, such as storm surge barriers

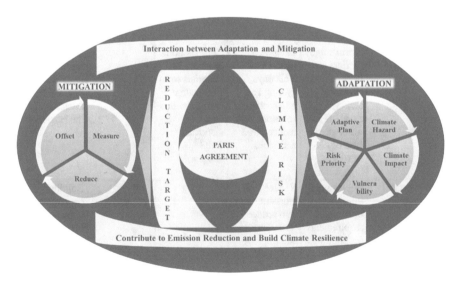

Fig. 5.7   Total carbon management model

Mitigation plan follows carbon management framework and to put into a simple term is to measure the CF, implement reduction programmes and offset the residual emissions. Building climate resilience, on the adaptation side, includes hazard identification, impact and vulnerability assessment, risk prioritization and adaptation plans implementation.

Finally, company also needs to assess the interaction relationship between different mitigation and adaptation measures to optimize them and incorporate them into business plan, policies and strategies to build overall physical, operational and financial sustainability.

References

Bowen F (2014) After greenwashing: symbolic corporate environmentalism and society. Queen Mary University of London. Cambridge University Press, Cambridge

BP TN (2018) BP TN qualifying explanatory statement for carbon neutral declaration for Castrol PCO engine oils and Castrol Engine Shampoo sold in Japan. BP Target Neutral. October 2018

BSI (2014) PAS 2060:2014 Specification for the demonstration of carbon neutrality. BSI. 30 April 2014.

BSR (2007) Who's going "carbon neutral"? A compilation by business for social responsibility. bsr.org

C40 (2011) Case study the world's first Carbon Neutral Sustainable City Dongtan, near Shanghai, China. C40 8 November 2011. https://www.c40.org/case_studies/the-worlds-first-carbon-neutral-sustainable-city. Accessed May 2019

C40 (2015) City climate hazard taxonomy: C40's classification of city-specific climate hazards. C40, Bloomberg Philanthropies and ARUP. New York

C40 (2018) C40 blog new toll will help cities understand interactions between mitigation and adaptation actions. https://www.c40.org/log_posts/new-tool-will-help-cities-understand-interactions-between-mitigation-and-adaptation-actions. Accessed May 2019

Cameron E, Harris S, Prattico E (2018) Resilient business, resilient world: a research framework for private-sector leadership on climate adaptation. Report. BSR, San Francisco

Carbon Footprint Ltd. (2017) Carbon footprint standard qualification requirements, Issue 1.1, 3 January 2017

Deutsch CH (2007) For Fiji water, a big list of green goals. New York Times, 7 November 2007. https://www.nytimes.com/2007/11/06/business/worldbusiness/06iht-water.4.8216299.html. Accessed May 2019

Environmental Leader, 2012. London Olympics generated 28% less CO2e than forecast. [Online] Available at: http://www.environmentalleader.com/2012/12/12/london-olympics-generated-28-less-co2e-than-forecast/

Gino F, Toffel MW, van Sice S (2013) FIJI water: carbon negative? Harvard Business Review, 611049-PDF-ENG, Published June 20, 2011 and revised on December 18, 2013

GIZ (2014) A framework for climate change vulnerability assessments. Published by Deutsche Gesellschaft für Internationale Zusammenarbeit (GIZ) GmbH, India Project on Climate Change Adaptation in Rural Areas of India (CCA RAI). ISBN 978-81-930074-0-2. September 2014

GreenBiz (2007) Fiji tries to green bottled water with its carbon-negative plan. GreenBiz. 7 November 2007. https://www.greenbiz.com/news/2007/11/07/fiji-tries-green-bottled-water-its-carbon-negative-plan. Accessed May 2019

Hammill A, Bizikova L, Dekens J and McCandless M (2013) Comparative analysis of climate change vulnerability assessments: lessons from Tunisia and Indonesia. Published by Deutsche Gesellschaft für Internationale Zusammenarbeit (GIZ) GmbH. March 2013

Hopwood A, Unerman J, Fries J (2010) Accounting for sustainability: practical insights, Published by Routledge, pp 184–185

HSBC (2008) HSBC and carbon neutrality. Issued by HSBC Holdings plc Group Corporate Sustainability. 8 Canada Square. London E14 5HQ. United Kingdom

ICO (2018) Carbon footprint methodology for Olympic games, by International Olympic Committee, released in December 2018

Inhabitat (2008) IS IT GREEN?: FIJI Bottled Water. Inhabitat. 25 September 2008. https://inhabi-tat.com/is-it-green-fiji-water/#more-14571. Accessed May 2019

IPCC (2012) Summary for policymakers. In Managing the Risks of Extreme Events and Disasters to advance climate change adaptation. A Special Report of Working Groups I and II of the Intergovernmental Panel on Climate Change. Cambridge University Press, Cambridge, UK and New York

IPCC (2014a) Climate change 2014: synthesis report. Contribution of Working Groups I, II and III to the Fifth Assessment Report of the Intergovernmental Panel on Climate Change. [Core Writing Team, Pachauri RK, Meyer LA (eds)]. IPCC, Geneva, Switzerland, 151 pp

Lewis D (2017) Energy positive: how Denmark's Samsø island switched to zero carbon. The Guadian. 23 February 2017. https://www.theguardian.com/sustainable-business/2017/feb/24/energy-positive-how-denmarks-sams-island-switched-to-zero-carbon. Accessed May 2019

London 2012 (2010) Carbon footprint study – methodology and reference footprint. Published December 2012. London Organising Committee of the Olympic Games and Paralympic Games Ltd (LOCOG). LOC2012/SUS/CP/0012

Mark and Spencer (2014) PAS 2060:2014 Specification for the demonstration of carbon neutrality. Qualifying Explanatory Statement in support of PAS 2060:2014 self-certification. Mark and Spencer Group Plc. May 2014

McGirk J (2015) Why eco-cities fail. Chinadialogue. 27 May 2015. https://www.chinadialogue.net/culture/7934-Why-eco-cities-fail/en. Accessed May 2019

Natural Capital Partners (2019) The carbon neutral protocol the global standard for carbon neutral programme, January 2019

Nottingham L, Yeo J (2017) How climate resilient is your company? Meeting a rising business imperative. Marsh & McLennan Companies' Global Risk Center, Marsh & McLennan Companies. 2017

O'right (2011). 碳中和承諾報告書, 400ml 綠茶洗髮精, Ver.1.0, 歐萊德國際股份有限公司, 2011年4月15日

SCS (2017) Qualifying Explanatory Statement for PAS 2060 Declaration of Achievement to Carbon Neutrality Prepared for: Planet Labs, SCS Global Services, August 2017

Teagarden MB (2017) Masdar city initiative: one step in the United Arab Emirates Journey to a New Energy Economy. Harvard Business Review. PRODUCT #: TB0479-PDF-ENG. Published 1st August 2017

ULI (2019) Climate risk and real estate investment decision-making. Urban Land Institute. 2019

USAID (2018) Designing climate vulnerability assessment, produced for review by the United States Agency for International Development. June 2018

World Development Movement (2013) HSBC and fossil fuel finance. Why the UK's biggest bank needs to pull out of coal. Media Briefing. October 2013

WorldGBC (2017) From thousands to billions – coordinated action towards 100% net zero carbon buildings by 2050

Chapter 6
Carbon Management Maturity Model

Carbon management on the mitigation side has been discussed in detail in Chap. 4, with an elaboration on the carbon management process from measurement to offsetting. Chap. 5 then introduced the model of total carbon management, combining the mitigation approach to achieve carbon neutrality and how business could build climate resilience from the adaptation side. When all these (i.e. mitigation and adaptation) happen within the organizational boundary of a business, we call it vertical carbon management, that is, to minimizing emissions and managing material climate risks within the organization.

In this chapter, we will extend the boundary from what company could control to its supply chain or whole value chain. As shown in Fig. 6.1, horizontal carbon management means company could extend the total carbon management into its internal and external processes and its products and engage stakeholders such as suppliers, customers, and communities to achieve truly overall emission reduction and to build the resilience.

Carbon Management Maturity Model (CM3) was first developed in 2007 by Supply Chain Consulting Pty Ltd., where I worked as a senior environmental consultant and market development manager during 2008–2009. Supply Chain Consulting acquired Viewlocity in 2006. Viewlocity was a Sydney-based company offering supply chain visibility and carbon management solutions, and its business was kept separate from Supply Chain Consulting's SAP practice in terms of its management team and accounting structure from day one of acquisition. So technically speaking and according to my best knowledge, Viewlocity originally developed CM3 and related carbon software solution CarbonView.

In April 2009, Fujitsu acquired Supply Chain Consulting Pty Ltd's core SAP business in Australia, Thailand and the Philippines, and its carbon emission tracking software business, Viewlocity, spun off as a separate Pty Ltd. company (Technology Transactions 2009). In October 2010, Viewlocity Holdings Pty Limited sold its CarbonView software solutions business and related assets to Element6 Group, Inc. (Eagle Corporate Advisers 2010).

© Springer Nature Switzerland AG 2020
S. W. W. Zhou, *Carbon Management for a Sustainable Environment*,
https://doi.org/10.1007/978-3-030-35062-8_6

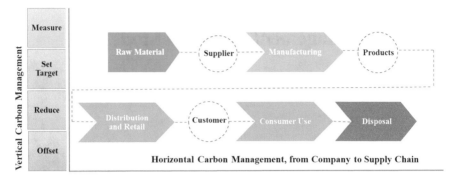

Fig. 6.1 Vertical and horizontal carbon management

In this chapter, I will first introduce the original CM3 as a tool to conduct horizontal carbon management for a business. I will then explain my modification to the existing model based on my consultancy experiences. A few cases would also be introduced to show how to use the model.

6.1   What Is CM3?

The Carbon Management Maturity Model provides a proven five-step process for company to implement its carbon management strategy, from simple responsive action to climate change to truly minimize the emissions, as shown in Fig. 6.2.

- Level 1 – The Basics

Company starts the carbon management journey by educating employees about the importance of implementing carbon management strategies, building initial awareness and developing an internal focus on improving basic operations. Different green programmes such as energy-saving workshops and waste recycling could be implemented at this stage, which gets the company ready for the carbon management.

- Level 2 – Company Level

Carbon footprint measured and emission reduced at the internal business level of the corporate. At this level, company focus on its own facility and operations.

- Level 3 – Process Level

Now company broaden its focus to include both internal and external processes across the extended supply chain. Company can even use carbon footprint information when making sourcing and production decisions.

- Level 4 – Product Level

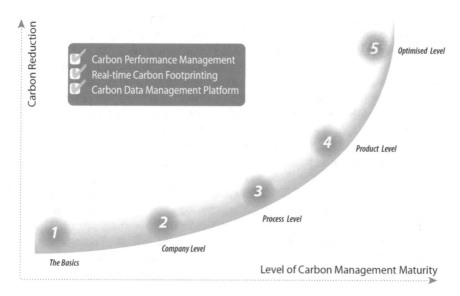

Fig. 6.2 The original Carbon Management Maturity Model. (Adopted from Supply Chain Consulting 2007)

Carbon footprint defined at product level. Company could allocate the process activity level carbon emissions to individual products to enable informed consumer activities.

- Level 5 – Optimized Level

Company performs analysis of financial and carbon relationship to balance emission reduction and financial performance. Integration of carbon and financial data in corporate decision-making to drive financially and environmentally sustainable business value improvements.

As seen from above explanation, CM3 provides a set of guidelines to determine where an organization is positioned in terms of environmental strategy and to create a long-term path to financial optimized carbon management practices.

It should be noted that no matter Supply Chain Consulting or Viewlocity, they are IT software companies. Therefore, the CM3 was initially developed in support for its carbon software solution development. CarbonView was originally developed by Viewlocity – one of the first purpose-built carbon management solutions, achieving its first sale in 2007. CarbonView is a carbon accounting and reporting software solution that enables users to calculate carbon emission, monitor it in real time and use intelligent algorithms when planning a balanced environmental and financially sustainable strategy to address GHG emissions. It consists of three integrated modules – Footprinter, Visibility and Optimiser – which, when implemented in support of CM3, combine to deliver a complete carbon management solution.

According to the original CM3 as above, the product-level CF measurement is not based on the traditional life cycle assessment, as a bottom-up approach, but the top-down carbon allocation approach, that is, measuring the CF at each life cycle stage by using facility-level CF measurement and then allocating the emissions to each process or activity of the product and sum them up to get the product CF. Carbon allocation method could give an estimated CF data for process and product for the business decision-making purpose, and compared with traditional life cycle assessment, it reduces the measurement time and cost.

In addition, CM3 does not consider the cost implications at each level. Carbon has a cost, no matter as defined by carbon trading or carbon tax as a regulatory means by the government as discussed in Chap. 3 or giving an internal carbon price to better reflect the cost-saving driver of carbon management as discussed in Sect. 4.1. Cost-effectiveness is also an important consideration in carbon reduction target setting as shown in Sect. 4.3.4. Therefore, financial implications of carbon as an asset should be embedded into each level of this model, instead of considering it at the optimized level 5. In other words, cost should be considered, and some level of optimization should be embedded at each level from level 1 to level 4.

6.2 Modified Carbon Management Maturity Model (*m*CM3)

In view of this, the modified Carbon Management Maturity Model (*m*CM3) as shown in Fig. 6.3 could be used in a more generic term. In this modified model *m*CM3, there are four levels: level 1, climate basics; level 2, company level; level 3,

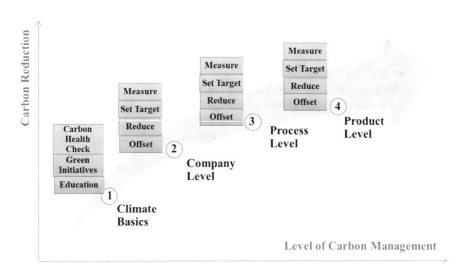

Fig. 6.3 Modified Carbon Management Maturity Model (*m*CM3)

process level; and level 4, product level. Except for level 1, carbon management framework or total carbon management (TCM) model could be applied at company level, process level and product level. As a result, *m*CM3 provides a framework or model that integrates both vertical and horizontal carbon management to company.

6.2.1 Climate Basics

This initial stage is the preparation stage, that is, to get the company ready for the carbon management journey. There are normally three different things need to be achieved at this stage:

- To conduct carbon health check to understand the company's current situation in terms of carbon management or where the company locates on the *m*CM3 diagram as shown in Fig. 6.3
- To educate internal stakeholders of the company, i.e. staff to let them understand the terminology of climate change, the impact and how the emission related to their daily life or business operations and how it is measured
- To implement related green initiatives or programmes, which are quick and easy to be accepted and implemented, such as the corporate sustainability initiatives or green programmes like installing recycling bins, pasting switch-off reminding stickers, purchasing energy-saving electrical and electronic appliance, signing up government or NGOs' climate campaigns and initiatives, etc.

Carbon Health Check

Figure 6.4 illustrates the carbon health check flow and the main topics of the checklist. To understand the carbon management status of the company, it is first to understand the carbon management drivers for the company, whether it has the strategy covered in its corporate sustainability or corporate social responsibility (CSR) policies, whether it is driven by the government regulations, whether it has been asked by its customers and whether the company has conducted different analysis to understand its potential cost savings opportunities.

The right side of Fig. 6.4 lists out the main topics along the carbon health check flow. For instance, if the company has carbon management strategies, detailed info under the topic CSR policy should be obtained. When company needs to report its CF to its government, on the checklist it needs further to know the assets owned by company, its operations and activities and how it manages its CF. If company needs to disclose its CF measurement and reduction to its client, we need to understand its products and services. Finally, we need to know if there have been any other information such as energy audit and supply chain management needed in the checkup. Table 6.1 gives an example of the carbon health survey checklist.

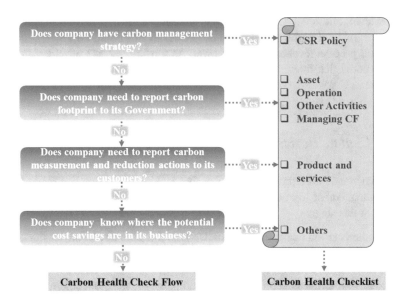

Fig. 6.4 Carbon health check flow

6.2.2 Company Level

Based on *m*CM3, company starts its real carbon management journey from its own operations first, that is, when company is ready with enough awareness and knowledge at level 1, it will then move up to level 2. Conducting carbon management at company level means measuring company-level CF, setting reduction target for the company, identifying mitigation measures and implementing them and offsetting the residual CF. The detailed process, tasks and deliverables have been discussed in Chap. 4, Sect. 4.7.

It should be noted that carbon management starts with the emission from company's direct activities, that is, emission within its organizational boundaries and direct emissions from Scope 1 plus purchased electricity and heat from Scope 2.

6.2.3 Process Level

A business process is defined as a specific event in a chain of structured business activities, the amalgam of which towards the final business goal. The process must involve clearly defined inputs and a single output. It can be categorized into management processes, operational processes and supporting business processes.

Examples of business processes include:

Table 6.1 An example of checklist used for carbon health check survey

Checklist	No	Yes	Data and comment
CSR policy			
Do you have a corporate social responsibility (CSR) policy?			
Do you have a current carbon management strategy in followings?			
CSR or corporate communication, marketing			
Developing new products and service			
Purchasing and supply chain management			
Investment planning			
Asset			
Does your company own a company car or car fleet?			
Does your company operate an industrialized AC			
Does your company generate electricity?			
Does your company have a boiler for internal heating?			
Operations			
What is the total number of light fittings and types of lights within your operation?			
How much electricity purchased annually?			
What are the annual fuel consumptions of petrol, diesel and gas?			
Types of waste generated and recycled?			
Other activities			
Employee business trips? Air? Train? Cars?			
Any outsource activities (transport, manufactory, services)?			
Does your company use contractors' vehicles, machineries?			
What materials and products purchased for your company?			
Managing carbon footprint			
Do you know current your carbon footprint?			
Do you know where the majority of your emissions arise from?			
Do you have a carbon reduction strategy?			
Any carbon reporting or carbon trading scheme at the moment?			
Products and services			
Have you done a life cycle analysis of your product?			
Do you know the amount of carbon released per product or service?			
Do you offer carbon neutral services?			
Others			
Have you conducted energy audit for your company?			
Do you want to include supply chain into carbon management?			

- Manufacturing – a production line, a product assembly process, a quality assurance process, a corrective/preventive maintenance process
- Logistic – shipping products, distribution of products, delivery of materials and wastes
- Finance – an invoicing process, a billing process, a risk management process

- HR – an orientation process, a leavers process, vacation request, updating employee data
- Procurement – purchasing raw materials, a procurement process of products and services

In mCM3, process-level carbon management means that process or business activities are not within company's four walls or its organizational boundaries but extend to its Scope 3 emissions. The owner of some activity data is not the company itself but its suppliers or contractors. For example, a production process includes input such as the raw material acquisition from suppliers and delivery of products by a logistic contractor.

Measuring process-level carbon footprint could let the company understand where and how emissions come from its designated business process, with a purpose to identify the opportunities to minimize it. Supply chain carbon measurement tools could be used here. For example, I used to use the database and analysis offered by Trucost, a UK-based company to report the "carbon hotspots" of suppliers. Such "hotspots" information helps to identify carbon-intensive suppliers and enable benchmarking suppliers on carbon performance. It also supports supplier engagement programmes for reduction target setting and how to achieve it. Company could also embed process-level carbon management into its existing supply chain management process or programmes.

Business Process Reengineering

Business process reengineering (BPR) is an approach to analyse and redesign company processes radically to obtain a specific business outcome. A goal of BPR is usually to analyse workflows and related tasks to optimize end-to-end processes and to eliminate tasks to achieve dramatic improvement in cost, quality and productivity.

There are a few successful cases of BPR in the 1990s. For instance, Ford used BPR and information technology to radically change its accounts' payable process to eliminate the invoice. In the new process, a buyer no longer needed to send a copy of the purchasing order form to the creditor administration. Instead, an order was registered and updated in the online database. As soon as the materials have been received, a warehouse man would update the materials received, and the payment is automatically made without waiting for the invoice to be received from the vendor. Through these changes in the business process, Ford had achieved a 75% reduction in employees in the administration department.

Taco Bell redesigned their business processes, focusing more on the retail service aspect and centralizing the manufacturing area. Ingredients of the food such as meat, corn shells, beans, lettuce, cheese and tomatoes for their restaurants were changed to be prepared in central commissaries outside the restaurant. At the restaurants, the prepared ingredients were then assembled when ordered. Better employee morale, increased quality control, fewer accidents and injuries, bigger savings and more time for focusing on customer business processes were some of the successes of the new way of work.

Hallmark used BPR to reduce product cycle from 3 years to 8 months, by creating a cross-functional team for product development based on the discovery that two-thirds of the product cycle was spent on planning and conceptualizing the card rather than on printing and production rework as had previously been thought.

There also have been a lot of BPR cases failed during implementation that could not achieve the original business goals. Habib in his paper reviewed and summarized the critical successful factor as well as causes of failure of BPR (Habib 2013).

Zinser et al. gave a best practice example of Porsche AG's development centre, who reduced the running time of the part procurement from 30 days to just 1 day by a specially developed 3x3 reengineering procedure (Zinser et al. 1998). The example illustrated a few critical success factors, such as a systematic procedure for the useful implementation of methods and tools and the establishment of a competent and motivated team leader and online process controlling supported by IT implementation.

BPR can also be undertaken with the strategic focus on green perspective; here is process-level carbon management. Take an example of the reengineering of the processes of a digital library published in a recent book (Unhelkar 2016). The physical search could be eliminated and replaced by an overall search system. Storage, reservation and loan processes are overlapped and could be designed as one central system to manage all these processes. The centralized system could be deployed into the cloud environment. As a result, various computing equipment can be replaced by a high-end energy-saving panel, and with all the changes, the overall carbon emission could be reduced. Lan estimated by process redesign a typical procurement process could reduce carbon emission by 44% (Lan 2011).

McKinsey Global Institute summarized 12 disruptive technologies that could have the potential to disrupt the status quo, alter the way people live and work, rearrange value pools and lead to entirely new products and services (McKinsey and Company 2013). It should be emphasized that these disruptive technologies, such as the pervasiveness of the Internet of Things (IoT) and artificial intelligence (AI), could facilitate companies to rethink their business processes and workflows radically. In the future, it is expected that BPR will continue to be part of the business digital transformation and technology-driven changes and it will play as an even more effective solution to the process-level carbon management.

6.2.4 Product Level

Product or service is the ultimate offering by the company, through which a business exists to generate its values, i.e. profits return to the shareholders, tax paid to the government, creating employment, developing skills of workers, driving innovation through research and development, making a contribution to the society, etc.

After company has minimized its environmental impacts from its own operations and has tried to manage its suppliers and contractors, the next level for company to move onto is the product level, where company could conduct a comprehensive assessment of its product or service, which reflects all the environmental impacts, here is carbon emission, through its life cycle stages.

Product CF measurement has been discussed in Chap. 2, Sect. 2.8. Product CF is reported in CO_2e for an assigned function unit of the product or service. It also needs to present the contribution of emission from each life cycle stage, where it gives the company an overview of the emission profile, and understand where to focus on reduction.

Ecodesign

Ecodesign can be understood as a process integrated within the design and development that aims to reduce environmental impacts and continually to improve the environmental performance of the products, throughout their life cycle from raw material extraction to end of life. ISO 14006:2011 provides guidance on how to incorporate environmental aspects into product design and development (ecodesign) processes within the framework of environment management systems (e.g. ISO 14001:2004) and quality management systems (e.g. ISO 9001:2008) (ISO 2011).

Ecodesign concept and framework could be used for product-level carbon management, which focuses on the integration of carbon footprint into product development and design, that is, when formulating the initial product concept, how to minimize the potential carbon emission of a product on ecosystem throughout its life cycle. Sanyé-Menguala et al. developed an ecodesign methodology which combined LCA of PCF, as shown in Fig. 6.5. PCF can be incorporated during the ecodesign process in different steps and with different purposes. First, PCF can be used as an environmental indicator in the quantitative assessment (Steps II and IV). Second, PCF can be used as strategy for environmental communication to the consumer (Step III) (Sanyé-Menguala et al. 2014).

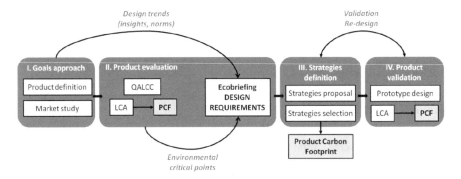

Fig. 6.5 Steps of ecodesign methodology and role of product carbon footprint (PCF). (Adopted from Sanyé-Menguala et al. 2014)

6.3   Case One: A Korean Manufacturer in China

6.3.1   Company Background

This company (let's call it Company A) is a Korean manufacturer who produced the handbag accessories for some of the world-brand handbags. Company A was located in Dongguan, Guangdong Provence of China. It was owned, managed and operated by Koreans. It had three factories near each other in Dongguan, one office buildings, one warehouse, several company cars and two dormitories mainly used for Korean management staff. The physical boundary is illustrated in Fig. 6.6.

When Company A was asked by some of its buyers from the UK on its carbon data in 2009, it had no idea what carbon was about. That was back in 2008, when carbon measurement and management were not well-recognized compared with today. Company A approached us at Supply Chain Consulting as a carbon consultant firm in Hong Kong.

6.3.2   Carbon Health Check

We paid a few visits to Company A in Dongguan. During our meetings, we have met a large ESH team who was responsible for environmental, safety and health issues within the company and understood that Company A had established extensive policies, systems and programmes on environmental and safety sides. Carbon concepts have been educated through our visits and meetings, and before this

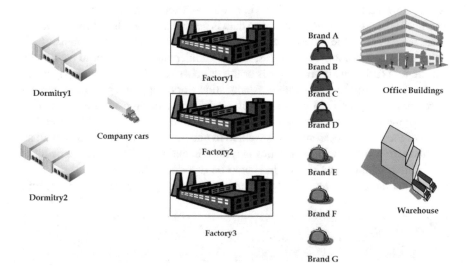

Fig. 6.6 Physical boundary of Company A

potential client could make their mind to undertake company carbon audit, carbon health check was proposed and conducted.

A carbon health check survey questionnaire was developed (based on the checklist framework as shown in Table 6.1) and used to understand Company A's current carbon awareness, environmental programmes and sustainability strategies. Scores were given to qualitative items, an example of which is shown in Fig. 6.7.

Quantitative data, such as electricity bills, the fuel consumption data for companies owned vehicles, water bills as well as the existing energy-saving installations and programmes have also been captured via the carbon health check questionnaire.

Based on this carbon health check exercise, it was then concluded that Company A was at level 1 of the *m*CM3 (Fig. 6.2) and was ready to move onto level 2 of the company-level carbon management.

6.3.3   Company Level Carbon Analysis

Company A did not commission us to start with a comprehensive carbon management, that is, measuring its CF and developing carbon reduction programmes and target immediately after the carbon health check. Instead they asked us to do a high-level assessment of its carbon emission. As I mentioned before, some quantitative data could be obtained from the carbon health check exercise, but that might not be sufficient to even get an estimation of the CF for the company. There were still some gaps to fill up at this stage.

On the one hand, the person in charge, in this case was the head of its EHS function as well as its management team, should understand what carbon measures are and how it relates to their business operation and activities, what their buyers are asking or what kind of carbon emission data is needed and for what purpose. On the other hand, we as the external consultant also should understand Company A's status quo, its operation, its business, its stakeholders and its market to try to help the client to define the business goals or the drivers to conduct carbon management at level 2. To fill up the gaps, awareness/education workshops on carbon management were arranged by its EHS head and delivered for its Korean executive team. This was also the preparation task to lead Company A to move from level 1 to level 2.

After the workshop, Company A wanted to understand its carbon impact with an intention to identify the opportunities to reduce it, and they also wanted to be more competitive with lower CF data that could be reported to their buyers. It was then agreed that more data were needed to conduct a company-level carbon assessment.

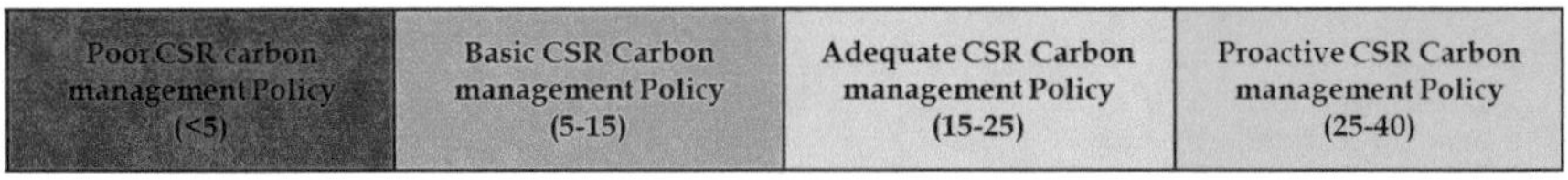

Poor CSR carbon management Policy (<5)	Basic CSR Carbon management Policy (5-15)	Adequate CSR Carbon management Policy (15-25)	Proactive CSR Carbon management Policy (25-40)

Fig. 6.7 Examples of scoring on qualitative items

Firstly, in order to know where emissions come from and how to reduce them, carbon emissions should be measured at the detailed level as much as possible. It means instead of the overall energy data for Company A, electricity consumption, refrigerant leakage, fuel, water and waste data should be collected at most detailed level, in this case was the entity or building level. Secondly, Company A wanted to report emission from staff commuting, as our client thought that compared with their main competitors in the USA, they would have a much lower emission from staff commuting as staff either lived in dormitories (Korean staff) or lived very near the factories that they normally took public transportation or motorcycles. Staff questionnaire survey was then rolled out to collect the transportation data. Lastly, Company A also wanted to report emission from waste disposal, as they thought their products were designed with minimized materials and they already implemented an impressive recycling and reusing initiative. Figure 6.8 shows the carbon emission analysis based on sources and entities, which was the major deliverable at this stage for Company A (data and proportion shown are not real for Company A, but for demonstration purpose only).

6.3.4   Product Carbon Labelling

Besides the emission profile and analysis, carbon reduction solutions were also suggested on the same carbon audit report delivered. In addition, as we knew the driver for Company A to conduct carbon management was to increase its competitive advantage with relatively lower CF, Company A was then recommended to move to product-level carbon management on level 4, where Company A could manage its product CF and disclose to their buyers. Carbon labelling was also proposed at this stage. But Company A did not take this forward. Back to 10 years ago, Company A was already a first mover and leader in carbon management in this part of the world.

6.4   Case Two: An International Hotel in Australia

6.4.1   Hotel X: Carbon Management Overview

Hotel X, located in Melbourne, Australia, is an international hotel, which belongs to an American hotel chain. Back in 2006, Hotel X was asked by its guests on its carbon strategies or how much to offset their stays at the hotel. In 2007, Australian government passed the National Greenhouse and Energy Reporting Act 2007 (NGER Act), under which the NGER scheme was established to require Australian corporate and facility (beyond a certain threshold) to report and disseminate company information about GHG emissions, energy production, energy consumption and other information specified under NGER legislation. In addition, Hotel X aimed

(a)

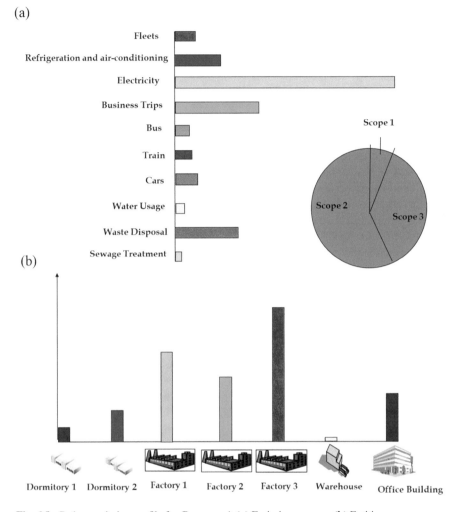

(b)

Fig. 6.8 Carbon emission profile for Company A (**a**) Emission sources, (**b**) Entities

to become a leader within the hotel industry on carbon management, so that it was proactively looking for solutions.

The business drivers for Hotel X to conduct carbon management could be summarized as follows:

- Government legislation (NGER)
- Demand from customers/guests
- "What is this hotel's Carbon Strategy?"
- "How much is it to offset my stay/conference?"
- Enhance hotel image
- Enhance credibility

- Develop an auditable and sound carbon management strategy
- Educate guests and staff of efficiency and carbon reductions

We were commissioned to use CarbonView as a tool to deliver a two-stage carbon management consulting job.

- Stage one – Company level

  - Facility (hotel) carbon footprint

- Stage two – Process level

  - Conference facility carbon footprint

    - What is the carbon cost of holding a conference/meeting at Hotel X
    - Carbon invoice presented to customer

  - Accommodation carbon footprint

    - What is the carbon cost of spending a night at Hotel X

      - The suite
      - The deluxe rooms
      - The standard rooms

    - Carbon invoice presented to customer

6.4.2 Stage One: Facility Carbon Management

To calculate the CF of the hotel itself, facility-level carbon measurement was conducted first, which was also the first step of company-level carbon management as level 2 on mCM3 model.

The physical boundary was the Hotel X itself, and the operational boundary included Scope 1, refrigerants and fuels; Scope 2, purchased electricity; and Scope 3, staff commuter, business travels and municipal solid waste disposal. Figure 6.9 shows the Hotel X's emission profile by scope and by emission sources (data and proportion shown are not real for Hotel X but for demonstration purpose only).

As a further analysis of CF at this stage, the emission for each department were presented in total carbon emission and carbon intensity, i.e. emission per employee, as shown in Fig. 6.10a, b, respectively. It was interesting to find that when the consultant presented the total CF for each department, there were no feedbacks or responses from the clients; but while the carbon intensity data was presented, it actually increased the staff morale in driving to reduce its CF. As such, some departments offered smart transportation card as an incentive to encourage its staff to take public transportation for the first mile or last mile, instead of driving all the way from home to office at hotel. It demonstrated a good example that carbon intensity in corporate context sometimes could be a better parameter to be reported

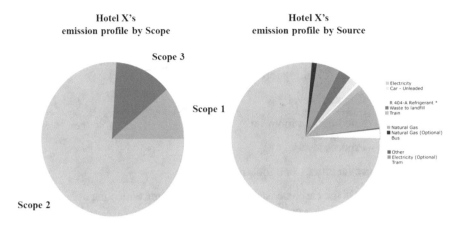

Fig. 6.9   Carbon emission profile for Hotel X

and presented for competition, benchmarking and driving positive changes to reduce emissions.

As the greatest component of the overall Hotel X footprint came from the electricity usage, as illustrated in Fig. 6.9, recommended mitigation options were then strongly linked to this emission source. Some recommendations are listed below:

• Switch to renewable energy

The GreenPower Program is a government managed scheme that enables Australian households and businesses to displace their electricity usage with certified renewable energy. By purchasing GreenPower, companies commit their GreenPower Providers to purchasing the equivalent amount of electricity from accredited renewable energy generators, which generate electricity from sources like wind, solar, water and bioenergy. Instead of recommending Hotel X to generate electricity from waste or renewable sources on-site, purchasing GreenPower was recommended to help Hotel X switch to part of RE electricity generation.

• Break down electricity circuit

Effective energy management requires detailed understanding on how electricity is being used. From the previous discussion, it was found that competition among different department was an effective way in driving changes. To install energy meters to break down energy consumption to the departmental level was then recommended.

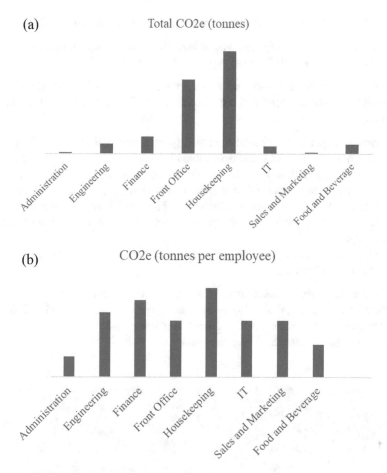

Fig. 6.10 (**a**) Total carbon emission and (**b**) carbon intensity for each department

6.4.3   *Stage Two: Conference and Hotel Room Carbon Management*

At this stage of the project, the key objectives are set by the project team as listed below:

- Create a "real" carbon invoice for actual conference activities.
  - Capturing carbon emission data while conferences are being held

- Create a "real" carbon invoice for one-night accommodation at the Hotel X.

  - Some in-room assumptions will be made based on average consumption rates
  - Will apply to a specific room type offered by the Hotel X

- Incorporate conference and hotel room carbon cost within financial invoices.
- Highlight activities open for mitigation relating to the conference and hotel room footprints.

Carbon management for Hotel X moved from company level 2 to process level 3. As illustrated in Fig. 6.11, the operational boundary of stage two was extended from stage one by adding more upstream and downstream emissions such as the emissions from raw material production, outsourced products and services, transportation of materials and products, etc. The GHG inventory for conference activities and one-night accommodation at Hotel X is shown in Fig. 6.12a, b, respectively.

The benefits obtained by Hotel X from this carbon management project can be summarized as below:

- A successful carbon conscious hotel working environment
- Decrease in carbon emissions since conducting carbon management in 2006/2007
- Decreased operational costs (e.g. energy and water costs)
- Improvement of hotel environmental image

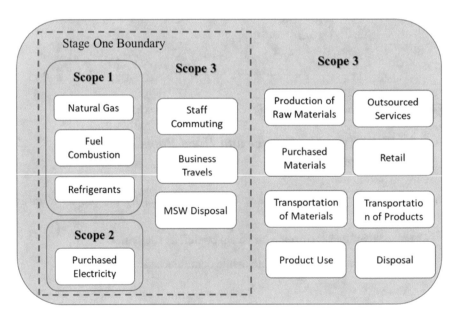

Fig. 6.11 Operational boundaries of stage two

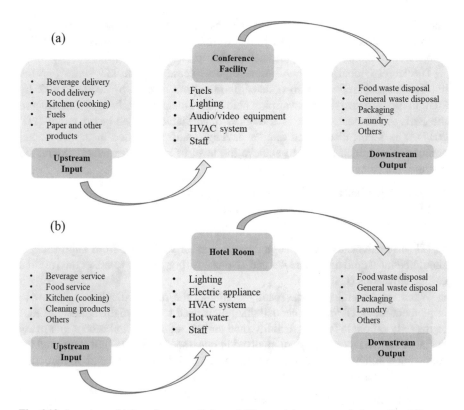

Fig. 6.12 Inventory of (**a**) conference activity and (**b**) one-night accommodation at Hotel X

6.4.4   Hospitality Sector Overview

The Hotel X project was taken place more than 10 years ago, which definitely made Hotel X a leader and first mover in terms of carbon management in hotel sector. The author would like to spend some time here to summarize the most up-to-date commitment and actions for the hotel industry, in the benefit to the readers to equip them with updated knowledge in conducting carbon management for hotels.

The hotel industry accounts for around 1% of global emissions. The International Tourism Partnership (ITP) is a global organization bringing together the world's most powerful hotel companies with 30,000 member hotels "embrace the ambition of science-based targets and encourage the wider industry to join their collaboration to develop carbon reductions at scale" according to its goals for 2030.

The Hotel Carbon Measurement Initiative (HCMI) is a methodology and tool which enables hotels to measure and report on carbon emissions in a consistent way. It was developed by the ITP and the World Travel & Tourism Council in partnership with KPMG and 23 global hotel companies. HCMI can be used by any hotel anywhere in the world, from small guesthouses to 5-star resorts. Over 24,000 hotels globally are using HCMI. Working with Greenview, ITP has launched the Hotel

Footprinting Tool which allows anyone easy access to the carbon and energy footprint of hotels worldwide. More information can be obtained from ITP's website: https://www.tourismpartnership.org/carbon-emissions/.

The Hotel Global Decarbonisation Report released by the ITP states that the hotel sector must reduce its absolute carbon footprint by 66% by 2030 and 90% by 2050 in order to keep global warming below the 2-degree threshold agreed upon in the Paris Agreement, which is a defined quantifiable science-based target (ITP 2017). The report also stated that "a paradigm shift is needed in the way hotel companies approach energy to achieve the level of reduction needed". Identified methods are categorized into:

1. Increasing efficiency of equipment and operations
2. Increasing the prevalence of renewable energy
3. Increasing "electrification"

ITP also advocated that Scope 3 emissions or emissions from supply chain should be measured and managed and also use carbon credits to offset the residual carbon emissions (ITP 2017).

When the technologies are readily available, the most difficult part is how to implement them. ITP recommended that hotels need to "evolve the mindset of ROI as the only rationale for sustainability and carbon reduction, and accept that many business drivers increasingly are involved in this issue" (ITP 2017). It suggested the following solutions to value energy and carbon and drive hotels' carbon reduction some of which have been discussed in Chap. 4 as well:

- Present the financial risk of inaction.
- By adding in the risk of cost increases for energy and managing carbon emissions, ROI calculations become more feasible.
- Set an internal price of carbon.
- Expand managerial accounting for carbon.
- Treat energy as a fixed-cost asset.
- Relate the impact of climate change on company growth strategy.
- Address climate change as CEO legacy opportunity.

6.5 Case Three: A Japanese Firm in Hong Kong

6.5.1 A Call and a Fax

This case started with a phone call received at Carbon Care Asia office in 2009/2010 when I was working as a senior consultant there. The caller was from a company called Firm P in Hong Kong and asked if we could offer the "carbon services" for them. When we wanted to further understand the company itself and what exactly they needed, the caller sent us a fax as shown in Fig. 6.13 and told us that it was what they wanted.

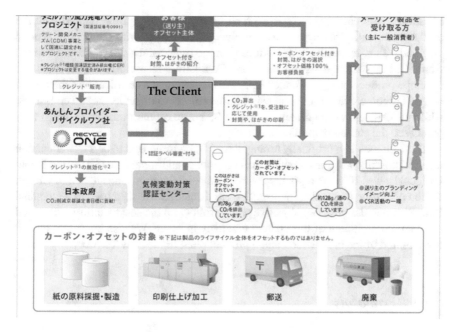

Fig. 6.13 The fax from Firm P

We lately found out from the website that Firm P is a subsidiary of the Japanese global printing company T, providing business forms printing, data management services and office machinery. Firm P located in Hong Kong is also Company T's regional headquarter of Asia Pacific. But what did they want in terms of "carbon services" or, in other words, what could we offer as a consultant? If it were you, what is your idea or what do you want to propose?

6.5.2 First Meeting

We did not rush into proposing carbon management consultancy service without further understanding our client's needs. Hence, we called for a meeting with our potential client for that purpose.

My sale colleague and I went to the first meeting, where the attendants from Firm P were their General Manager and his assistant. Neither my colleague nor I could read or speak Japanese while both the manager and the assistant were Japanese, but luckily the assistant could understand a little Mandarin. No matter how difficult in communicating due to the language barriers, we still got the useful info from this meeting.

Back to the fax as shown in Fig. 6.13. It was from Firm P's parent company T in Japan. The Manager of Firm P would like to understand what it meant. Of course, it is not the Japanese, which I did not understand, but the content and the concept behind. As you can see, there is a graph of supply chain on the bottom, and the products shown in this figure are two types of envelopes with CF data of around 78 g and 128 g, respectively. In addition, on the right-hand side, you could find the word "CSR". We made a good guess that Company T measured the PCF of its two envelope products via LCA under Japanese Government's certification scheme and bought carbon credits from some wind farm projects to offset its carbon emissions, while the cost of carbon offsetting would be borne by their customers when they purchased the carbon neutral envelopes.

This first meeting, originally arranged for 1 hour, turned out to be a free education lecture for 2.5 hours, during which the concepts of anthropogenic GHGs, the Kyoto Protocol (it was 10 years ago), the business drivers of carbon management, PCF and carbon offsetting and carbon neutrality were discussed. We then tried to understand Firm P's potential business objective was to follow up the sustainability (CSR used by them) commitment of their parent company in Japan. We also knew that Firm P had a rented office in Kowloon and a product plant in the New Territory.

6.5.3   Proposals

Following our meeting, based on the *mCM3* framework, the following has been proposed to Firm P as summarized in Table 6.2. It should be noted that our proposal is focused on the production plant of Firm P. Hence, instead of starting the company level carbon management for the whole company, we chose the facility-level carbon management for plant first.

As shown in Fig. 6.14, although they were same type of products (i.e. envelopes in this example), they had a bit different processes in their life cycle stages. One type of envelope was produced at the plant, with paper as its raw materials, while the other type of envelope was purchased directly from the suppliers. The other processes such as printing, delivery of the final products and disposal after use were

Table 6.2 Proposed carbon management services for Firm P

Level	Carbon management services	Marketing claims
Facility level	Carbon audit for production plant in the New Territory Developing carbon footprint management plan Carbon offsetting	Carbon neutral plant
Process level	Carbon measurement for its printing process within the production plant, plus the materials used Developing carbon footprint management plan Carbon offsetting	Carbon neutral printing services
Product level	Measurement of PCF of envelope Developing carbon footprint management plan Carbon offsetting	Carbon neutral products
Company level	Carbon audit for the Firm P	Carbon emission reporting and disclosure

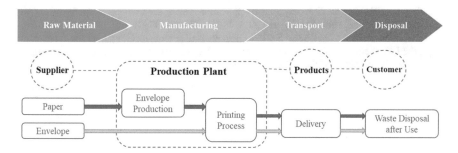

Fig. 6.14 Firm P's process and its supply chain

all the same. In order to measure PCF of two envelopes, different suppliers would be engaged to obtain the raw material footprint data, and the emissions of the plant should be allocated to different activities for different products as well.

Firm P did not commission us to conduct carbon management afterwards, as the company later figured out that unlike Japan, as Japan ratified Kyoto Protocol, Hong Kong did not have an obligation nor a mandate in carbon reporting or reduction. In other words, the company could not find their own business drivers, or value proposition of carbon management, which could bring them a justified business case. What we have done in the end was the advocacy, awareness education or further capacity building, free of charge, for our potential clients, as majority of time we did, as a consultant.

References

Eagle Corporate Advisers (2010) CarbonView Software – sale of assets to Element6 Group, Inc 31 October 2010. http://www.eagle-advisers.com/carbonview.php. Accessed May 2019

Habib MN (2013) Understanding critical success and failure factors of business process reengineering. Int Rev Manag Bus Res 2(1):1–10

ISO (2011) Environmental management systems — guidelines for incorporating ecodesign, 1st edn. International Organization for Standardization, Geneva

ITP (2017) Hotel global decarbonisation report. Aligning the sector with the Paris Climate Agreement towards 2030 and 2050. International Tourism Partnership, November 2017

Lan Y (2011) Reengineering a green business. Int J Green Comput 2(1):1–11

McKinsey & Company (2013) Disruptive technologies: advances that will transform life, business, and the global economy. McKinsey Global Institute, Washington, DC

Sanyé-Menguala E, Lozanoa RG, Farrenya R, Oliver-Solàa J, Gasola CM, Rieradevalla J (2014) Chapter 1: introduction to the eco-design methodology and the role of product carbon footprint. In: Muthu SS (ed) Assessment of carbon footprint in different industrial sectors, vol 1. Springer, Singapore

Supply Chain Consulting (2007) Mastering proactive carbon management – a 5-step model to achieving a green supply chain. Supply Chain Consulting. 2007.

Technology Transactions (2009) Fujitsu Australia acquires supply chain consulting for A$48m. 29 April 2009. http://tmt-transactions.com/fujitsu-australia-acquires-supply-chain-consulting-for-a48m/. Accessed May 2019

Unhelkar B (2016) Green IT strategies and applications: using environmental intelligence. CRC
    Press, Boca Raton. Published 22 June 2011. ISBN 9781439837801
Zinser S, Baumgärtner A, Walliser F-S (1998) Best practice in reengineering: a successful example
    of the Porsche research and development center. Bus Process Manag J 4(2):154–167. https://
    doi.org/10.1108/14637159810212325

Chapter 7
Carbon Reduction Solutions

Based on a recent McKinsey report (McKinsey 2019), global primary energy demand plateaus after 2035 despite strong population and economic growth and electricity consumption doubles until 2050, while renewables are projected to contribute over 50% of generation by 2035. In addition, global carbon emissions are projected to peak in 2024 and decrease by around 20% by 2050 due to a relatively rapid phase-out of coal in the power sector (primarily driven by a reduction of demand in the Chinese power sector). However, this outlook remains far from a 2-degree pathway, as defined by the median of all scenarios in IPCC as shown in Fig. 7.1.

There are four scenarios on this graph. RCP2.6 represents a scenario to keep global warming likely below 2 °C above pre-industrial temperatures. It means emission has to peak just in the next few years and heads down to zero. That's what IPCC put forwards as a path consistent with a "likely" achievement, meaning two-thirds probability of staying below 2 °C. But our aim is not just to keep warming "likely" below 2 °C, but well below it. A steeper path to get to zero emission faster than RCP2.6 is needed as we need to get the zero emission in the second half of this century as fast as possible.

Now the gap becomes even wider – on the one hand, McKinsey report shows we are far away from IPCC's 2-degree pathway (McKinsey 2019), and on the other hand, even IPCC's defined pathways are not enough. In this chapter, I will discuss how we could close this gap. I will first introduce the "deep decarbonization" concept proposed by UN Sustainable Development Solutions Network (SDSN) and some results from country's case studies. Then I will move to the business sector. Different low-carbon solutions for buildings, transportation and waste management will be introduced. I hope that could lay out ideas and technological solutions to the students and give insights to consultants and companies when planning carbon reduction initiatives.

© Springer Nature Switzerland AG 2020
S. W. W. Zhou, *Carbon Management for a Sustainable Environment*,
https://doi.org/10.1007/978-3-030-35062-8_7

7.1   Deep Decarbonization Pathways

Deep decarbonization is a process to get the emissions down to zero within the next few decades. Convened under the Institute for Sustainable Development and International Relations (IDDRI) and the Sustainable Development Solutions Network (SDSN), the Deep Decarbonization Pathways Project (DDPP) is a collaborative global research initiative to understand what needs to be done for individual countries' low-carbon economy transition and to approach net zero GHG emission to consist with the internationally agreed goal of limiting anthropogenic warming to less than 2 °C (DDPP 2015).

7.1.1   The Three Pillars of Deep Decarbonization

The keys to the deep decarbonization transition are what referred by DDPP as the three pillars, covered by all the 16 participating countries in DDPP and under all the scenarios under analysis. The three pillars are energy efficiency, decarbonization of electricity and fuel switching, as shown in Fig. 7.2.

First, energy efficiency in DDPP is defined as energy intensity per GDP. It means if the country could get more economic output per unit of energy, it could also get more economic output per unit of carbon dioxide. For a given size of the economy, the country is trying to use less of energy and emit less of carbon. Second, decarbonizing the electricity system means electricity is generated not by coal-fired or natural gas-fired power plants, but by renewable energy such as solar, wind, hydropower or geothermal power, where no additional emissions are contributed by electricity generation. Lastly, fuel switching means switching energy end-users to low-carbon supplies instead of changing the primary energy source. Examples of fuel switching could be driving from gasoline-powered internal combustion engine cars to electric vehicles, heating from boilers to electric heat pumps, etc.

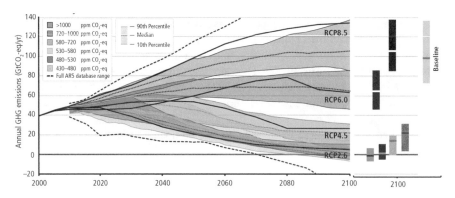

Fig. 7.1 GHG emission pathways 2000–2100: all AR5 scenarios. (Adapted from IPCC (2014a, b))

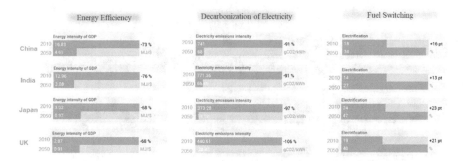

Fig. 7.2 Three pillars of decarbonization for different countries. (Data from DDPP Website deep-decarbonization.org)

Participating countries in DDPP have their different ways of implementing these three pillars with their country-specific strategies and available technology. Deep decarbonization pathways for the USA, Australia and China will be introduced below. US case demonstrates the energy system transformation, and in Australia case, it aims to illustrate how to use three pillars of decarbonization strategies in building sector. China, as a developing economy, would present another different pathway.

It has to be emphasized that deep decarbonization could not be achieved in absence of one or two of the three pillars, as they interact with each other. Taking electric vehicles as an example, increase in population and GDP may result in increased vehicle sales and the total mileage of driving, but electrification of vehicles itself could not drive deep decarbonization of road transport, if electricity used to power electric batteries is still generated from dirty coals.

7.1.2   Case Study: US Deep Decarbonization

In US DDPP report (Williams et al. 2014), it was concluded that an 80% reduction in GHG emissions by 2050 in the USA was an achievable result and there were multiple feasible technology pathways. It examined four different scenarios: High Renewables Case, the High Nuclear Case, the High Carbon Capture and Storage Case, and the Mixed Case. These different cases not only referred to the different kinds of electricity system, but they also explored different kinds of transportation systems, different measures in industry and different kinds of building measures and others. All four of these cases were able to achieve deep decarbonization in the USA without hardships and consistent with realistic constraints about grid reliability and infrastructure inertia (Williams et al. 2014).

One potential low-carbon transition (i.e. mixed case scenario) of the US energy system is illustrated in the Sankey diagrams as shown in Fig. 7.3. Sankey diagrams use arrows to represent the major flows of energy from supply side to end-use side, with the width of the arrows being proportional to the magnitude of the flows.

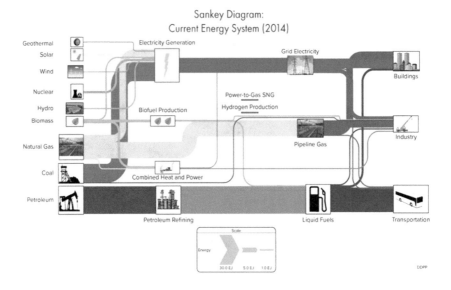

Fig. 7.3 Sankey diagram for US energy system. (Adapted from US DDPP Report (Williams et al. 2014))

As shown in Fig. 7.3, US current primary energy system in 2014 was dominated by very large amounts of fossil fuels, like natural gas, coal and petroleum, and very small amounts of renewable energy, like geothermal, solar, wind and hydro. Moving from the left-hand to the right-hand side of this diagram, it can be found that energy was consumed by three mega end-use sectors buildings, industry and transportation, and in the middle the blue line on top was electricity generation, where showed different energy sources going into the grid and how grid electricity was coming out to different sectors.

Now moving down to the deeply decarbonized energy system in 2050 in Fig. 7.3, it can be found that as primary energy source, renewable energies would be much larger and the fossil fuel lines would be much smaller, where coal would not exist and petroleum and natural gas would be greatly reduced. At the same time, the blue line that represented grid electricity would become thicker, which means that while more electricity would be generated to fulfil the economic growth and people's increased living quality in 2050, low-carbon electricity generation and electrification of energy consumption would achieve a deeply decarbonized energy system.

7.1.3 Case Study: Australia Deep Decarbonization

In addition to the three pillars defined for all countries in DDPP, Australia has a fourth one, that is, reducing non-energy emissions in industry and agriculture. Based on Australia's DDPP report (Denis et al. 2014), the deep decarbonization pillars for Australia are listed below:

1. Energy efficiency: Greatly improved energy efficiency in all energy end-use sectors, including passenger and goods transportation, residential and commercial buildings and industry.
2. Low-carbon electricity: Replacement of existing fossil fuel-based generation with renewable energy (e.g. hydro, wind, solar and geothermal), nuclear power and/or fossil fuels (coal, gas) with carbon capture and storage (CCS).
3. Electrification and fuel switching: Switching end-use energy supplies from highly carbon-intensive fossil fuels to lower-carbon fuels, including low-carbon electricity, hydrogen, sustainable biomass or lower-carbon fossil fuels.
4. Non-energy emissions: These emissions can be reduced through process improvements, material substitution, best practice farming and implementation of CCS.

As shown in Fig. 7.4, Australia's current emission was 17 tCO$_2$/capita in 2012, and it could be reduced by more than 80% to 3.0 tCO$_2$/capita in 2050. It could be further reduced to 1.6 tCO$_2$/capita if emissions from the production of exports were excluded. Australia has substantial potential to offset emissions via land sector sequestration. It was concluded by its DDPP report that Australia can achieve net zero emissions by 2050, using technologies that exist today, while maintaining economic prosperity (Denis et al. 2014).

Let's have a deep look at decarbonization pathway at building sector. Buildings contribute around 30% of Australia's GHG emissions, and buildings could reach net zero by 2050 or earlier with good technologies, right policies and best practices and behavioural changes.

First, energy efficiency. Energy use in buildings needs to be cut half for both residential (energy use per household) and commercial buildings (energy use per square

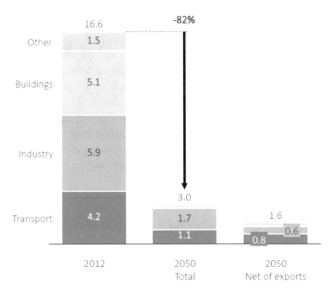

Fig. 7.4 Australia's pathway on energy reduction (per sector per capita). (Adapted from Australia DDPP Report (Denis et al. 2014))

meter) by 2050, and current technologies such as energy-efficient lighting, energy-efficient air conditioning, smart building shelf, etc. show that this target could be achieved through ensuring that new buildings adopt the green standards and replace equipment by best practice models at the end of its useful life.

Second, low-carbon electricity. Electricity used in building can come from on-site installations such as PV panels or BIPVs or from off-site renewable energy generations or from the grid electricity, which is a clean one.

Last, electrification and fuel switching. A switch from natural gas to a decarbonized electricity supply results in near elimination of emissions from buildings by 2050. This involves a move from gas to electricity for all heating, hot water and cooking equipment.

Overall, modelling as part of the Australia DDPP showed that emissions from residential and commercial buildings will reduce by 97% between 2012 and 2050, even with a substantial increase in the number of households and size and output of the commercial sector, if the above three pillars were used: energy efficiency, renewable energy, and electrification (ClimateWorks Australia et al. 2014).

What are significant barriers to achieving net zero in buildings? These are cost (high premium of energy-efficient fixtures and renewable energy installation); knowledge (building owners or property management have little knowledge on the most updated solutions and technologies that they need capacity building on how to move to zero emission); motivation; and the last but most important the lack of policy, as the good policy could not only motivate the industry zero carbon movement but also provide incentives for that transition. The policy framework for deep decarbonization in building sectors should then include the following:

- Pathway to net zero 2050.
- Interim targets for 2030 and 2040 to set stepwise target and to provide a framework to monitor the progress.
- Set up a government office which coordinates with all related government departments and oversees the zero carbon transition.
- Mandate building energy-efficiency standards and construction code.
- Implement policies to incentivize the sector, such as feed-in-tariff for on-site renewable energy installation, tax deduction or grant for green buildings.

7.1.4 Case Study: China Deep Decarbonization

In 2007, China surpassed the USA as the world's largest carbon emitter, and as of 2016 China accounted for about 28% of world fossil fuel CO_2 emissions. As a developing economy, although China would shift from high-speed growth to moderately high growth, with more people moving from rural areas to cities, the ongoing urbanization will increase the demand for infrastructure, building materials and consumer goods. It is anticipated that China will experience an increase in energy use and carbon emissions from such economic development in near future. At the moment, China's energy generation was still dominated by coal (as seen in Fig. 7.5), which not only resulted in carbon emission but also in severe air pollution. Hence China needs to transit to low-carbon economy, not only to tackle the climate change issues but also to mitigate the serious pollution problems and at the same time to meet its further development needs.

Based on China DDPP report (Teng et al. 2015), China's DDPP pathway is analysed using Strategy Analysis on Climate Change in China model (SACC) developed by National Centre for Climate Change Strategy and International Cooperation (NCSC). SACC is a bottom-up model, including an accounting model for end-user sectors including industry, transportation and the building sector, and a least-cost optimization method for the power generation sector with the aim of analysing technology options and relevant carbon emissions (Liu et al. 2016). The SACC model uses 2010 as a base year and is designed to study short-term (before 2020), medium-term (2020–2030) and long-term (2030–2050) China's energy generation, energy demands and carbon.

Considering China's increasing energy demand, decarbonization of the power sector is a crucial component of any pathway for the achievement of low-carbon development. As shown in Fig. 7.5, based on one scenario for the deep decarbonization pathway of China (Teng et al. 2015), carbon intensity of power generation will decrease more than 90% in 2050, from 741 gCO_2/kWh in 2010 to 68 gCO_2/kWh. Non-fossil electricity will dominate electricity production, with the ratio climbing

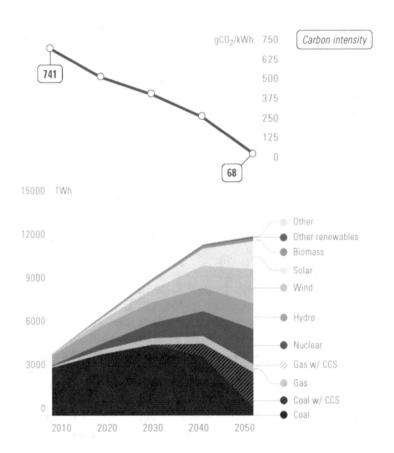

Fig. 7.5 One scenario of deep decarbonization of carbon intensity for China. (Adapted from China DDPP Report (Teng et al. 2015))

to 72%, which is from less than 20% in 2010 to 34% in 2020 and 43% in 2030. By 2050, wind power will be responsible for 26.8% of total non-fossil power generation, and solar power will contribute 21.7% of it. CCUS serves as an important measure to reduce emissions from the power sector after 2030.

In addition, Table 7.1 summarizes the final energy consumption and total carbon emission for different end-user sectors. The figures were calculated based on China DDPP report (Teng et al. 2015). Total final energy use was projected to increase 162% in 2050, with an increase of 40%, 92% and 130% for industry, building and transportation sector, respectively. Nevertheless, total carbon emission was expected to reduce by 36% by 2050 via electrification of end uses and increasing the ratio of non-fossil fuels in the overall energy mix. By 2050 the industry sector was expected to remain as the largest contributor to the emission. Emission from industry and building would decrease by about 52% and 30%, respectively, from 2010 levels, while emission from the transportation sector would increase by 67% over the same period.

Table 7.1 Final energy consumption and total emission for China, by sector

	Industrial	Building	Transportation	Total
Energy demand, Mtoe				
2010	1,233	333	267	1,833
2050	1,709	632	627	2,968
Change (%)	+40%	+92%	+130%	+162%
Total carbon emission, $MtCO_2$				
2010	5,758	1,580	807	8,145
2050	2,745	1,106	1,350	5,201
Change (%)	−52%	−30%	+67%	−36%

7.2 Solutions for Business

The above section discussed the deep decarbonization strategies for countries, which mainly focused on energy efficiency, low-carbon electricity and electrification and fuel switching, to reach net zero emission by mid of this century. Now the question is: "What can we do as a business or each individual?"

Prof Jem Bendell strongly argued that we are in the face of "an inevitable near-term social collapse due to climate change" that we really need to reassess and rethink our work and our life (Bendell 2018). Instead of the top-down strategies from national level, what would be the bottom-up approach? The answer is that we need disruptive technology and social innovation in a rapid way.

Entrepreneur Paul Hawken's book *Drawdown* (Hawken 2017) paved us a more optimistic view in addressing climate change. *Drawdown* describes the 100 most substantive solutions to global warming, which is the product of years of research by different scientists, researchers, students, etc. by mapping, measuring and modelling each solution to climate change. For each solution, the carbon impact, its cost and savings, how it works and how to adopt it have been investigated, discussed and communicated. It is also very important that all solutions modelled are already in place, well understood, analysed based on peer-reviewed science and expanding around the world.

What are the solutions to business sectors or individuals who have no control over how electricity or material and food is generated, or how land is used, but could make wiser choice when using the electricity or other types of energy, products and materials. Solutions are categorized into seven sectors in *Drawdown* (Hawken 2017): electricity generation, food, women and girls, buildings and cities, lad use, transport and materials. To give the readers a good sense of the possible solutions in this book that could be applied in business sector or even in our daily life, the author selected the top ten ranked solutions based on carbon reduction as summarized in Table 7.2.

Among these solutions, I would like to divide them into two categories: one is those business could adopt by themselves, and the other is those business could support. Let's have a look at the top ten solutions. Take, for example, refrigeration – as we know that the refrigerants normally have a high GWP. While CFCs and HCFCs have already been banned due to their high ozone-depletion potential, based on Montreal Protocol, the common replacement HFCs have a GWP from 1,000 to 9,000. Choosing a refrigerant with lower GWP and higher efficiency is the solution that company could adopt. Solutions such as onshore wind, solar farms, tropical forests and silvopasture might be difficult for company to adopt or implement within their business operation or scope, but company could choose to support them, for instance, by green financing wind project, by purchasing green power from solar farm, by donating to protect the tropical forests in Amazon, or by livestock products from silvopasture managed pasture.

It should be noted that educating girls and family planning are two of the top ten solutions in reducing global carbon emission. It might not be practical to many developed economies or business in these economies where females have already got high education. But according to *Drawdown*, these two solutions influence family size and global population, which is one of the most powerful levers available for emission reduction by curbing population growth. Women with higher education have fewer birth rates and are more resilient and capable of facing climate change impacts. More and more highly educated women in developed countries become the stewards of environmental protection, organic food, green farming, reduced consumerisms, and upcycling, which is effective in promoting a low-carbon life style.

Lastly, technology and innovation cannot be considered in isolation but need to be put in the broader socio-economic system with a complex interplay between markets and various institutional actors: innovators who champion new sustainable technologies, investors who see market opportunities in these sustainable technologies, executives who steer large organizations towards profitable and sustainable opportunities, customers who are willing to pay for them, activists who advocate businesses to invest in green innovation, and governments who incentivize new technologies through regulation, taxes and other policy levers.

Table 7.2 Top ten decarbonization solutions from drawback.org

Ranking	Solution	Sector	GHG reduced (gigatons)	Net cost (billions USD)	Net savings (billions USD)
1	Refrigeration	Materials/building	89.74	N/A	−902.77
2	Onshore wind	Energy	84.60	1,225.37	7,425.00
3	Reducing food waste	Food	70.53	N/A	N/A
4	Plant-rich diet	Food	66.11	N/A	N/A
5	Tropical forests	Land use	61.23	N/A	N/A
6	Educating girls	Women	59.60	N/A	N/A
7	Family planning	Women	59.60	N/A	N/A
8	Solar farms	Energy	36.90	−80.60	5,023.84
9	Silvopasture	Food	31.19	41.59	699.37
10	Rooftop solar	Energy/building	24.60	453.14	3,457.63

7.3   Low-Carbon Buildings

Low-carbon buildings are buildings which are specifically designed, operated and maintained with GHG reduction in mind. Although there is no carbon emission threshold to define it, a low-carbon building is a building which emits significantly less GHG throughout its life cycle when benchmarking with the same type of buildings.

In this session, drivers for low-carbon buildings will be discussed first, and different solutions under passive design and active design will then be introduced. Next, how to operate and manage a low-carbon building will be talked. Marketing tools such as green building labelling schemes will be further introduced. And lastly the future development of super-low-carbon buildings or net zero buildings will be explored.

7.3.1   Drivers for Low-Carbon Buildings

Low-Carbon Buildings
Figure 7.6 illustrates the detailed life cycle stages of a building, which includes design a building, raw materials extraction and production, transportation of construction materials and products to the construction site, on-site construction and installation, building operation and maintenance, retrofitting and renovation after

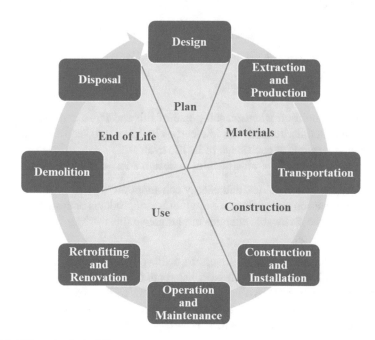

Fig. 7.6 Life cycle of a building

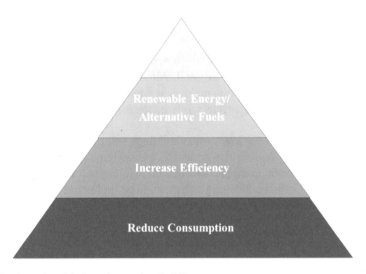

Fig. 7.7 Hierarchy of design a low-carbon building

Fig. 7.8 Top 32 drivers for green buildings

years of use, demolition when reaching end of life and reuse and disposal of the construction materials. To achieve a low-carbon building is to keep the carbon emission as low as possible at each stage of its life cycle.

Figure 7.7 shows the overall strategies in design a low-carbon building:

1. To reduce the materials' use and energy consumption
2. To use low-carbon materials and energy-efficient products or systems
3. To switch to the renewable energies or alternative fuels
4. To offset emission

Then what are the business drivers to achieve a low-carbon building? Hong Kong Polytechnic University professors listed out 64 drivers for green buildings in total in a recent paper (Darko et al. 2017). Top 32 of these drivers as shown in Fig. 7.8 are classified into five categories: external, corporate level, project level, property level and individual level. It is easy to agree that legislation is the top driver for change and adopting green building standard. Regulations themselves often vary according to country or region, and international regulatory requirements that act as vehicles for change towards sustainability have been developed, among which the most important initiatives such as Paris Agreement and UN Principle for Responsible Investment (PRI) will be discussed in detail.

Cost savings such as energy conservation and reduced whole life cycle cost at property level and reduced construction cost at project level and reducing environmental impact such as environmental protection and water and resources conservation at property level and waste reduction at project level are also top drivers. Incentive schemes also play an important role. Green building labelling scheme and its relevant incentive schemes will be discussed later.

In a recent study (Murtagh et al. 2016), 28 architectural designers in 14 small firms in the London area were interviewed, and it was found that the first driver for most architectural designers to implement sustainability was client demand.

Based on Fig. 7.8, it also should be noted that 11 out of 32 top green building drivers are on corporate level, which means internal corporate sustainability drives the adoption of green buildings from various areas such as marketing benefits, CSR, high return on investment, competitive advantages and images, etc.

Readers may ask what the difference between green buildings and low-carbon buildings is. In author's opinion, there is no fundamental difference between the two, and most of the time, people use both terms interchangeably. Nevertheless, green building is defined by the green building labelling schemes, and low-carbon building normally uses carbon emission as the parameter to evaluate the building's performance, both of which include energy, electricity consumption, material and waste, water consumption, indoor environmental quality, etc.

Driver #1 Paris Agreement

Paris Agreement as a global collective treaty has been discussed in Chap. 1. How does it drive building sector? Building construction and operations accounted for 36% of global final energy use and 39% of energy-related carbon emissions in 2017, and energy use per m^2 in buildings needs to be reduced by 30% by 2030 to be in line with the Paris Agreement (UNEP 2018). By 2060, 230 billion m^2 of additional buildings will be constructed (UNEP 2017). Hence, ambitious action is needed without delay to reform the buildings and construction sector's energy performance, to avoid locking in long-lived, inefficient building assets for decades to come and to keep the Paris Agreement on track.

Market tools such as building carbon trading could also provide incentives for carbon reduction in buildings, for instance, Tokyo cap-and-trade, the world first scheme including office buildings, launched in April 2010 and aimed to cut emissions by 25% by 2020 from 2000 levels, which has already been achieved in 2014.

When we discussed climate risks in Chap. 5, we know that climate risks are sector specific, and both physical risks and transition risks associated with climate change have financial impact to the buildings. Market intelligence provider Four Twenty Seven and real estate technology company GeoPhy have partnered to assess the exposure to the physical impacts of climate change of 73,500 properties owned

Fig. 7.9 How green building can contribute towards meeting the SDGs. (From worldgbc.org)

by 350 listed real estate investment trusts (REITs). It was found that the most exposed REITs (defined as REITs that hold sites where half or more of the sites are exposed to one of the climate hazards, i.e. sea level rise, hurricanes and typhoons, flood, water stress, heat stress) were primarily geographically concentrated in Asia – Japan, Hong Kong and Singapore in particular. Champion REIT stands out due to the geographic concentration of its properties in Hong Kong and the city's high exposure to flood, sea level rise and typhoons (Four Twenty Seven 2018). Therefore, to design, build and operate a climate-resilient low-carbon building is a must under the scenario of Paris Agreement.

Driver #2 SDGs

As we discussed in Chap. 1, in the same year of Paris Conference, United Nations set 17 goals with 169 targets covering social and economic development issues, serving the agenda for 2030, called "the Future we want" – a harmonious future with no poverty, zero hunger, good health, quality education, clean water and air, equality, peace and justice and so on and so forth. Implementation of SDGs is to localize the SDGs to highlight the role of regional government, sectors and each individual. As effort of the building sector, World Green Building Council (WorldGBC) has conducted a series of mapping on how green buildings, such as green home and green office, could provide the foundations to meet 9 of 17 SDGs as shown in Fig. 7.9. In other words, SDGs or the future we want is one of the major drivers for green buildings.

Driver #3 UN-PRI

From investment point of view, the responsible investment is an approach to incorporate ESG factors into investment decisions, to better manage risk and to generate sustainable, long-term return. United Nations-supported Principles for Responsible Investment, in short PRI, launched a guide to help investors – both asset owners and investment managers – who are implementing ESG integration techniques in their investment process. PRI's *Impact Investing Market Map* identified six environmental themes and four social themes, and green building is one of the environmental thematic investments. Based on PRI, green buildings are defined as "companies that generate their revenues from buildings designed, constructed, operated, maintained, renovated and destroyed using environmentally-friendly and resource-efficient processes", and to obtain green building certification and to manage ESG issues are essential in terms of responsible investment (PRI 2018).

In one of PRI's reports on its signatories' practices (PRI 2016), in building sector, of 114 investors, 85% thought that ESG factors have helped them to identify risks and opportunities, while 60% reported ESG considerations have had an impact on the prices offered or paid for property. It showed that buildings could get "green premiums" with high ESG performance or green building certifications. It also reported that almost 50% of respondents have abandoned potential investments because of ESG issues. Ninety-six percent of the 47 respondents stated the financial impact of green buildings was positive.

Driver #4 Disclosure
Carbon reporting and disclosure has been discussed in Sect. 2.7.2. All the reporting and disclosure requirements, such as ESG reporting for listed companies, carbon disclosure and climate-related financial disclosure, are all applied to buildings. To manage building's environmental and carbon performance and to report them is another important driver for green buildings or low-carbon buildings.

7.3.2 Active and Passive Design

The main environmental aspects in buildings are energy consumption, water use, materials use and waste generation, among which energy consumption is normally the major contributor to carbon emission. In this section different approaches and solutions will be discussed to reduce energy use in buildings.

Passive Design

> Passive design is to reduce the demand of mechanical and electrical provision by utilizing material and geographic characteristics and natural resources to satisfy environmental quality while minimizing energy consumption.

Building passive design is architectural design that fully utilizes the natural resources, takes advantage of climate to maintain comfortable and healthy indoor environment and could achieve a lifetime thermal comfort, low life cycle energy cost and carbon reduction. Passive design solutions in solar heat gain, natural lighting, natural ventilation and green roofs will be introduced.

Solar Heat Gain

Controlling the solar heat gain during summer time and heat loss during the winter time is an effective way to reduce the energy demand for cooling and heating in the buildings. Heat will be exchanged through building roofs, walls, windows or glazing. By using sun path analysis, the following could be considered:

- To decide the building's orientation, for example, long facade with windows facing north-south.
- To choose low emissivity materials, such as low-e glass curtain walls, or solar reflectance of internal furnishes.
- To install external shadows, such as sun-shading, balcony windows or overhangs. One innovative solution worth to mention here is Al Bahar Towers responsive facade, designed by Aedas. Its Mashrabiya shading system was computer-controlled to respond to optimal solar and light conditions.

Natural Lighting

Natural lighting could not only reduce the energy demand from the electric lighting but could also benefit the occupant's health. The WELL Building Standard (WELL), managed and administered by the International WELL Building Institute (IWBI), is a performance-based system for measuring, certifying and monitoring features of the built environment that impact human health and well-being. WELL sets thresholds for indoor sunlight exposure to encourage natural lighting to support circadian and psychological health.

Daylight could be achieved through skylights in atriums like in many shopping malls or open windows or corridors like in schools and libraries. Indoor lighting level can be reduced by photo-cell diming sensors. Daylight can also be harvested by light tubes or light pipes. A good example is Sing Yin Secondary School, the recipient of the "Greenest School on Earth 2013", where a heliostat was fitted on the rooftop to track the movement of the sun and collect sunlight and direct it via solar optical fibre to the floor below to illuminate a physics laboratory, which substitutes part of the electric lighting to save electricity and reduce carbon emissions.

LightEFX, an Australian solution provider to the architectural and construction industry utilizing fibre optic lighting and natural daylighting technologies, compared the different natural lighting systems and declared that their patented Sunportal technology allows the best natural daylight to reach and illuminate any indoor environment, into deep building spaces (compared with effective transmission distance up to 20 m for light tubes and 30 m for optical fibres), and because it uses the latest IR-cut coating technology, there is no heat loss or gain associated with the transmitted daylight so the cost of heating or air conditioning and its greenhouse impact is also reduced. More information can be found from its website https://www.lightefx.com.au/.

Natural Ventilation

Natural ventilation uses outdoor air or wind as free energy source to satisfy occupancy's thermal comfort. Natural ventilation could reduce the dependency on mechanical and electrical ventilation systems, which could reduce the capital cost and the maintenance and replacement cost. But there are also some drawbacks. My office used to be located at Hong Kong Science Park, where windows of our office could be opened. To save energy, majority of time, we opened the window to use natural wind, but a lot of times, it was too noisy, as there were a few construction sites nearby. Air conditioning must be used when it is too hot outside or the air quality is too bad or the climate is too humid, as natural ventilation could not provide enough cooling comfort at extremely hot days, unable to filter the outdoor air and control humidity.

Whether or not an occupant is thermally comfortable is generally assessed as a heat balance of the occupant in connection with their surroundings. As defined by ASHRAE Standard 55 (2010), there are six primary factors that must be addressed when defining conditions for thermal comfort:

1. Metabolic rate
2. Clothing insulation
3. Air temperature
4. Radiant temperature
5. Air speed
6. Humidity

Natural ventilation is normally not provided in a mechanically ventilated building, as there would be concerns if windows are open and at the same time the air conditioning is switched on. Such circumstances might happen when the occupants forget to close the window, or the occupants are too greedy to have maximal comfort. Nevertheless, operable vents could be considered strategically and designed to the windows and mechanical ventilation systems. For instance, one feature of Hysan Place, which achieved the Platinum Rating of Final Assessment under the BEAM Plus NB V1.1 of the Hong Kong Green Building Council on 7 June 2013, was the operable vents. During right external conditions, occupants can open top and low operable vents along perimeter zones of the office floors; then the combined wind and stack effect will enable occupants to enjoy natural ventilation. Mixed mode ventilation combines natural ventilation (via operable vents) and mechanical ventilation/air conditioning to provide and enhance occupants' comfort at office floors. Occupants may enjoy the natural ventilation in transition season or non-office hours according to their own comfort and ventilation requirements. Once the vents are opened, unnecessary air conditioning to individual perimeter zone will be automatically switched off. The effective area benefited from natural ventilation would be a 5 m deep space behind the facade. Free Cooling System or Economize Cycle uses oversized air handling units and air ducts.

Green Roof

With rapid urbanization, or at highly populated metropolitans like Hong Kong and Singapore, urban heat island (UHI) effect means higher temperature compared with rural areas that may impact people's thermal comfort and health. Major causes of UHI are listed as follows:

- Albedo

    - Albedo for concrete which absorbs more radiation
    - Albedo for vegetation

- Removal of vegetation and water

    - Less evapotranspiration
    - Less evaporation

- Low building permeability

    - Stagnant air

Green walls and green roofs could be effective approaches to mitigate UHI and improve the urban microclimate, by reducing direct solar radiation, reducing surface temperature and increasing evaporative cooling. Besides protecting buildings against solar radiation and temperature fluctuations, green roof could also reduce building's energy demand on space condition by direct shading (Morau et al. 2012). A study conducted by Dr. Sam Hui at the University of Hong Kong showed that green roofs can significantly moderate the daily temperature fluctuations with maximum temperature attenuation (i.e. daily temperature fluctuation of bare roof versus that of green roof) of 9.8–18.4 °C for three green roofs under study. It also estimated that the U-value of the roof would be reduced by around 40% and the corresponding saving of annual total building electrical energy (for top floor only) would be up to 4.6% (Hui 2009).

Active Design

> To satisfy environmental quality with minimum energy consumption of mechanical and electrical provision through efficiency improvement.

Active design uses or produces energy to maintain a comfortable and healthy environment in the building. Active design normally contains mechanical devices such as fans, air conditioning, electric lighting systems, lifts, pumps, etc. and transport the absorbed energy to other locations within the buildings.

HVAC

Based on Hong Kong Electrical Mechanical Service Department's (EMSD) report, the air conditioning uses 35% and 29% of electricity in residential and commercial sectors, respectively (EMSD 2018). Air conditioning is the largest electricity end-user. Passive design tries to reduce the cooling load for the buildings. Electricity can also be saved by using more efficient chillers like water-cooled chiller, oil-free chiller, variable speed drive and district cooling.

Water-cooled air conditioning systems (WACS) using freshwater cooling towers for individual building is generally more energy-efficient, consuming up to

20% less electricity than air-cooled air conditioning systems. But large amount of fresh water is required, so that when doing sustainability report, we normally encounter a good electricity reduction due to chiller's retrofitting from air-cooled to water-cooled, but at the same time, we find a significant increase in water consumption. In Hong Kong, seawater is commonly used by the once-through water cooling system, especially for large centralized air condition system along the seafront of Victoria Harbour, to take away heat of condenser and eventually discharge into the sea.

In chillers, lubricant normally deposits on to chiller, which will reduce chiller efficiency. Oil-free chiller eliminates lubricant by adopting magnetic bearing to increase its overall efficient.

Variable speed drives (VSDs) can cut a chiller's annual energy use by up to 30% while maintaining operating reliability over a wide range of condition. The compressor speed can be reduced to more closely match the load than a constant speed chiller at part-load when cooling capacity is reduced.

District cooling system (DCS) is a centralized cooling system which provides chilled water to the air conditioning system of numerous user buildings via underground chilled water pipe network. DCS is an energy-efficient air conditioning system as it consumes 35% and 20% less electricity as compared with traditional air-cooled air conditioning systems and individual WACS using freshwater cooling towers, respectively. More benefits of DCS can be found from EMSD's website https://www.emsd.gov.hk.

Lighting

Lighting accounts for 15% of electricity consumption for commercial energy end-users in Hong Kong, which makes lighting the second largest electricity consumer after air conditioning (EMSD 2018). To install energy-efficient light bulbs is often treated as the "low-hanging fruit" in energy management for low-carbon buildings, with low capital cost, short return period and high SIR. Figure 7.10 illustrates the journey of energy-efficient lighting technologies.

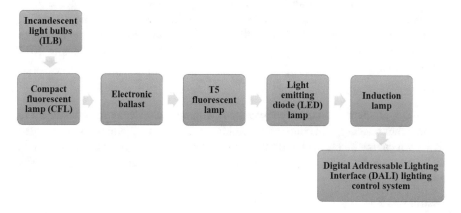

Fig. 7.10 Energy-efficient lighting technologies

Table 7.3 Comparison of different lighting technologies

Technology	CRI	Efficacy (lumen/W)	Lifetime (hours)
ILBs	98–100	12–18	750–2,500
CFLs	80–90	60–70	6,000–10,000
T5	70–90	75–100	15,000–24,000
White LED	70–90	60–90	25,000–50,000

Incandescent light bulbs (ILBs) have been replaced by compact fluorescent lamps (CFLs) since the 1980s. CFLs are compactly designed fluorescent tubes with their ballast and gas-filled tubes assembled together, which save around 75% of electricity compared to incandescent lamps and last up to 10,000 hours, i.e. six to eight times longer than typical ILBs (Liu et al. 2013). Solid-state circuit of electronic ballast appeared in the market in the early 1980s to replace the electromagnetic transformer. As electronic ballast contains no copper windings, energy losses could be reduced and it could also start the CFLs more quickly, producing less flicker to extend its lifetime. Since the 1990s T5 fluorescent lighting has emerged and has now taken over the T8 fluorescent lighting as a proven and cost-effective option for general lighting (Liu et al. 2013). Light emitting diode (LED) is now becoming the most commonly used energy-saving lighting solution for commercial buildings.

There are three parameters used to evaluate the lighting technologies: luminous efficacy (lumens per Watt), expected lifetime (hours) and colour rendering index (CRI). Table 7.3 summarizes the lighting performances of different technologies as a reference, based on US EPA's Energy Star (energystar.gov) and HK EMSD data (Liu et al. 2013):

In addition to the lamps, lighting control has been proven to reduce lighting energy consumption in commercial and industrial buildings by switching off the lighting when not needed. Technologies include occupancy sensors, daylight sensors, time scheduling, bi-level switching, dimming technology and intelligent system such as Digital and Addressable Lighting Interface (DALI) system (Liu et al. 2013). Smart lighting systems include LED lighting technologies, advanced sensors and universal communication interfaces to work in IoT ecosystem (Higuera et al. 2018). Light itself can be used for wireless communication, the technology of which is called Li-Fi. The idea of Li-Fi that utilizes light to transmit data between devices was first introduced by University of Edinburgh's professor Harald Haas in 2011 (Haas 2011).

Electrical Appliance

Energy-efficient appliances are energy-saving ones chosen by the consumers to use in either residential or commercial buildings. Different governments have issued their own energy labels to classify the energy-saving levels of the products and help the consumers to make wise decision. Table 7.4 summarizes the different energy labels as a reference.

Distributed Energy

Distributed energy, also called decentralized generation or on-site generation, is small-scale units of local generation connected to the grid at distribution level.

Table 7.4 Comparison of different energy labels for electrical appliance

Government	Name of the label	Products covered	Rating system
Australia	Energy Rating	Refrigerators, freezers, air conditioners, dishwashers, dryers, washing machines, televisions, computer monitors, three-phase air conditioners, swimming pool pumps	Energy consumption (generally kWh/year), 1–10 stars (10 most efficient) (includes half stars)
Canada	EnerGuide Program	Air conditioners (room), freezers, refrigerators, dryers, washing machines, combination washer/dryers, dishwashers, ranges/ovens, wine chillers	Labels display the energy (kWh/year) used and how this compares with the lowest and highest energy consumption for similar products. Air conditioner ratings are based on the energy efficiency ratio (EER) of the unit
EU	Come On Labels	Washing machines, dishwashers, refrigerators, air conditioning units, televisions, dryers and light sources	Energy (kWh/year or per cycle), efficiency rating A–G (A most efficient), although new label scales generally show a highest rating of A+++ with the lowest rating of D; the visible end scales depend on the product
Hong Kong	Mandatory Energy Efficiency Labelling Scheme (MEELS)	Air conditioners, refrigerators, CFLs, washing machines, dehumidifiers	Energy (kWh/year), efficiency rating (grade) 5 to 1 (1 most efficient)
USA	Energy Star	Originally only computers, monitors and printers, but has now been expanded to cover a wide variety of appliances, equipment, building products and even homes and windows	The label indicates that the product is among the most efficient of its type, either because it is in the top percentile of the range on the market or because it exceeds the Energy Performance Standards level by a specified margin

Distributed energy is a low-carbon solution for buildings or community. Firstly, because its scale is small, it could deploy renewable energy such as rooftop solar PV panels, natural gas turbines, microturbines, wind turbines, biomass generators or energy from waste plants to have cleaner power generation. Secondly, it is more reliable that it could improve the energy security and makes business more climate resilient, especially under more frequent extreme weather that has a threat on blackout resulted from power generation, transmission or distribution. Thirdly, it is more efficient. Compared with conventional central generation, less waste will be lost and the application of distributed cooling and heating also increases the efficiency. Lastly, new technologies on energy storage, smart meters and inverters, as well as the demand from electrification such as EV chargers, enable the distributed energy to become a disruptive solution in this and coming decade. The recent report published by Arup highlighted that from 2020 to 2030, the electricity system

became increasingly decentralized that low-carbon distributed and micro generation will contribute nearly half of the UK's generation capacity (Arup 2018).

7.3.3   Building Management and Operation

Low-carbon building management and operation is to maintain both passive and
active systems as described above and at the same time to manage energy, water,
resources and waste. *ISO 50001:2018 Energy management systems – Requirements
with guidance for use* gives organizations a recognized framework for developing
an effective energy management system, which helps organizations to develop and
implement an energy policy, to set achievable targets for energy use and to design
action plans to reach them and measure progress. This might include implementing
new energy-efficient technologies, reducing energy waste or improving current processes to cut energy costs. Two commonly used tools energy audit and retro-
commissioning will be introduced and discussed.

Energy Audit
Defined by Energy Audit Code 2015 (EMSD 2015), an energy audit is "the systematic review of the energy consuming equipment/systems in a building to identify
energy management opportunities (EMO), which provides useful information for
the building owner to decide on and implement the energy saving measures for
environmental consideration and economic benefits". An energy audit starts with
collecting the building energy consumption data and relevant information; then
reviewing and analysing the collected information and the performances of existing
equipment, systems and installations and comparing with performances at relevant
energy-efficient modes of operation; and finally identifying of areas of energy inef-
ficiency and the means for improvement (EMSD 2015).

Retro-commissioning
Based on EMSD's *Retro-commissioning Technical Guidelines on Retro-
commissioning 2017*, retro-commissioning (RCx) means a knowledge-based systematic process to periodically check an existing building's performance to identify
operational improvements that can save energy and thus lower energy bills and
improve indoor environment. RCx aims to resume system efficiency to design standard and to optimize the building services installations operating efficiency. RCx
is about:

- Finding out how efficient the system/equipment is operating
- Identifying how the system/equipment can operate at better/more efficient condi-
  tions (energy-saving opportunities (ESOs))
- Carrying out improvements
- Measuring and verifying savings
- Maintaining improved operation methods

Figure 7.11 compares energy audit and RCx. It is said that an energy audit is
like a snapshot of a building, while RCx is like a video recording of the building.

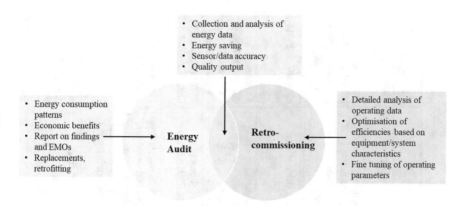

Fig. 7.11 Comparison between energy audit and retro-commissioning

Energy audits focus on the EMOs, which normally involve new capital investment on new equipment replacement and retrofitting. RCx focuses on repairing, recalibrating and reprogramming devices, investment cost of which is low, but it can extend the equipment life and also enhance the operation team's knowledge and skills.

7.3.4 Green Building Labelling Schemes

Green building certification and rating system is one of very important external drivers for low-carbon buildings. It not only provides the green building standards and benchmarks but could also provide financial incentives or mandate with appropriate policy instrument. For instance, in Hong Kong, new government buildings with a construction floor area of more than 5,000 m² have to attain "Gold" rating or above under BEAM Plus (Hong Kong's Green Building Label scheme). As a prominent financial incentive to private development, since April 2011, certification by BEAM Plus is one of the prerequisites for the granting of gross floor area (GFA) concessions for certain green and amenity features in development projects. Table 7.5 summarizes and compares popular green building rating schemes.

7.3.5 Low-Carbon Building Development

WorldGBC Advancing Net Zero

In its report *From Thousands to Billions – Coordinated Action towards 100% Net Zero Carbon Buildings By 2050,* WorldGBC calls for a dramatic and ambitious transformation towards a completely zero carbon built environment, through the dual goals of:

Table 7.5 Comparison of green building labelling schemes

Scheme	Country	Managing organization	Categories	Assessment criteria	Award/rating
BEAM	Hong Kong	HKGBC	New buildings; existing buildings; neighbourhood; interiors	Site; materials; energy use; water use; IEQ; innovation	Platinum; Gold; Silver; Bronze
BREEAM	UK	BRE	New construction; in-use Refurbishment and fit-out; communities	Energy; health and well-being; transport; water; materials; waste; land use and ecology management; pollution No prerequisites for in-use	Outstanding; Excellent; Very good; Good; Pass; Unclassified
CASBEE	Japan	JSBC (Japan Sustainable Building Consortium)	Pre-design; new construction; existing building; renovation	Energy efficiency; resource efficiency; local environment; indoor environment	Rank C (poor), rank B−, rank B+, rank A, and S (excellent)
Green Globes	Canada/US	Green Building Initiative in the US BOMA Canada	New construction/significant renovations; commercial interiors	Energy; indoor environment; site; water; resources; emissions; project/ environmental management No prerequisites	5 Globes: 4 Globes; 3 Globes; 3 Globes: 2 Globes: 1 Globe
Green Mark	Singapore	Building and Construction Authority (BCA)	New buildings; existing buildings; beyond buildings; within buildings	Energy efficiency; water efficiency; environmental protection; IEQ; other green and innovative features	Platinum; Gold^{plus}; Gold; Certified
LEED	US	U.S. Green Building Council	New construction (NC); existing buildings, operations and maintenance (EB O&M); commercial interiors (CI); core and shell (CS); schools (SCH); retail; healthcare (HC); homes; neighbourhood development (ND)	Sustainable sites; water efficiency; energy and atmosphere; materials and resources; IEQ; locations and linkages; awareness and education; innovation in design; regional priority through a set of prerequisites and credits	Platinum; Gold; Silver; Certified
Three Star	China	Ministry of Housing Urban-Rural Development	Residential; commercial; public	Land; material; energy; water; IEQ; operation	★★★ ★★ ★

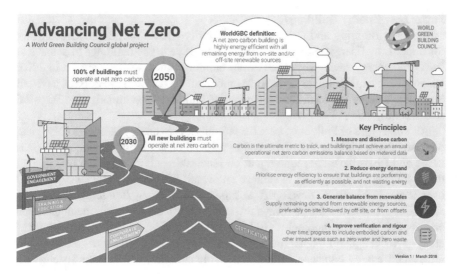

Fig. 7.12   The pathways to net zero carbon buildings

- All new buildings must operate at net zero carbon from 2030
- 100% of buildings must operate at net zero carbon by 2050

Advancing Net Zero is a global campaign by WorldGBC to accelerate uptake of net zero carbon buildings to 100% by 2050. As discussed in Sect. 5.1, WorldGBC thinks building could achieve net zero by energy efficiency measures and energy supply from renewable sources on-site or off-site. Figure 7.12 illustrates the WorldGBC's infographic on pathways to net zero carbon buildings.

Based on its key principles, there are five stages of commitment, which provides a framework for organizations to develop globally ambitious yet locally relevant, flexible and universally viable solutions for their portfolio to both reduce energy demand and achieve net zero carbon emissions:

1. Commit: Advanced trajectory for all new and existing buildings within direct control of the organization to operate at net zero carbon by 2030; regulate for all buildings to operate at net zero by 2050.
2. Disclose: Measure, disclose and assess annual asset and portfolio energy demand and carbon emissions.
3. Act: Develop and implement a decarbonization roadmap outlining key actions and milestones towards energy efficiency and renewable energy.
4. Verify: Demonstrate enhanced energy performance, reduced carbon emissions and progress towards net zero carbon assets and portfolio.
5. Advocate: Act as a catalyst through core business activities for further action within respective supply chains.

Currently there are 23 business organizations, 23 ties and 6 regions that have signed this commitment. More information could be found from its webpage: https://www.worldgbc.org/thecommitment.

Super Low Energy Buildings

Singapore's Building and Construction Authority (BCA) introduced Green Mark for Super Low Energy (SLE) rating in 2018 to drive office buildings to achieve at least 60% energy saving over the current building codes, which were set in 2005, which means SLE buildings cannot use more than 100 kWh/m²/year. BCA gave out Singapore's definition on different low-carbon buildings as summarized in Table 7.6 (BCA 2018).

All three types of buildings, i.e. SLE buildings, zero energy buildings (ZEB) and positive energy buildings (PEB), need to achieve 60% energy efficiency, whereas ZEBs and PEBs have the requirement on renewable energy supplies. In doing all these, Singapore is targeting to achieve an overall 80% energy efficiency by 2030 over 2005 level. BCA developed a *Super Low Energy Buildings Technology Roadmap* that charted the pathways towards SLE via development, demonstration and application of technologies (BCA 2018). As shown in Fig. 7.13, four focus areas have been identified and prioritized, namely, passive strategies, active strategies, smart energy management and renewable energies, which is consistent with what we discussed so far.

Table 7.6 Definition of different low-carbon buildings in Singapore

Category	Energy efficiency (beyond Green Mark Platinum)	Renewable energy (RE) vs. energy consumption (EC)	Building types
Positive energy building (PEB)	60% energy savings over 2005 level	RE ≥ 110% EC	Public schools; camps; landed houses, etc.
Zero energy building (ZEB)		RE ≥ 100% EC	Mid-size office buildings: institute of high learning buildings, etc.
Super low energy building (SLEB)		–	High-rise commercial buildings etc.

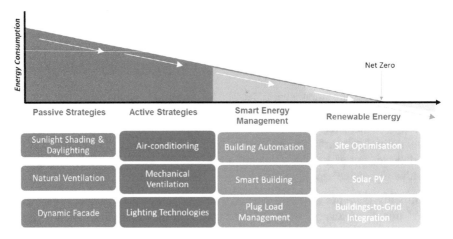

Fig. 7.13 Energy reduction strategies to achieve SLE buildings (BCA 2018)

C40 – Net Zero Building Declaration

At city level, C40s Net Zero Building Declaration was launched to demonstrate cities' commitments to slash emissions by 2030 and to build thriving, healthy and equitable communities by pledging to enact regulations and/or planning policy to ensure new buildings operate at net zero carbon by 2030 and all buildings by 2050, which matches WorldGBC's targets.

To meet this commitment, signatory cities will:

- Establish a roadmap for the commitment to reach net zero carbon buildings
- Develop a suite of supporting incentives and programmes
- Report annually on progress towards meeting the targets and evaluate the feasibility of reporting on emissions beyond operational carbon (such as refrigerants)

7.4 Smart Transportation

7.4.1 Emissions from Transportation

Transportation is the movement of people and goods by buses, cars, trucks, trains, ships, airplanes and other vehicles. It is and will continue to be a critical component for human mobility and economic growth. The majority of greenhouse gas emissions from transportation are carbon dioxide emissions resulting from the combustion of fossil fuels, like diesel and gasoline, in internal combustion engines. Figure 7.14 shows the contribution of emission from transportation (IEA 2018). Globally, transport emission contributes 24% of the total carbon emissions from fuel combustion. Compared with different regions, Asia has only 14% which might be due to its compact city environment and most areas in developing stage. Compared with different countries, transport emission contributes most in New Zealand, i.e. 48%, while other developed countries like the USA, UK and Canada are in the range of 30–35%; Asian developed economies like Hong Kong, Japan and South Korea are also similar to each other, with a 17–18% contribution of transport emission to total emission from fuel combustion. Least contribution of transport emission of total emissions from fuel combustion is in China, which is only 9% at the moment.

In the USA, transport has already become the economic sector which contributes largest carbon emissions. In China, the car ownership was still 58 vehicles per 1,000 persons in 2010, which was a tiny fraction compared to 804 per 1,000 in the USA. However, China's annual vehicle sales have grown quickly from 1.4 million vehicles in 1994 to 18 million in 2010 and surpassed the USA and all other countries in vehicle sales. It was projected that China's vehicle population would increase by 13–17% per year, reaching more than 500 million by 2030 (Wang et al. 2011).

Before introducing the mitigation solutions, first we need to understand how carbon emission could be assessed from transportation. Based on IPCC report on transport (IPCC 2014a, b), for each mode of transport, direct carbon emissions can be decomposed into:

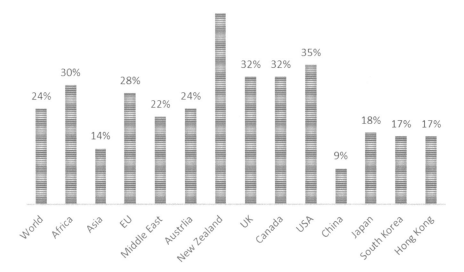

Fig. 7.14 Carbon emission from transport (% of total emission from fuel combustion)

- Activity – total passenger-km/year or freight tonne-km/year
- Transport infrastructure and modal choice
- Energy intensity – directly related to vehicle and engine design efficiency, driver behaviour during operation and usage patterns
- Fuel carbon intensity – varies for different transport fuels including electricity and hydrogen

As more than 72% of global transport emission is from road transport, which includes cars and vans, freight trucks, buses and motorcycles (IPCC 2014a, b), this book will only focus on low-carbon solutions to road transport. Based on the above discussion and the method from King Review (King 2007), the carbon emission from road transport could be assessed in the following Eq. (7.1):

$$\text{Transoport emission} = \sum_{\text{Mode}} \left[\text{Activity data} \times \text{fuel CO}_2 \text{ efficiency} \times \text{vehicle efficiency} \right] \tag{7.1}$$

Based on this equation, the mitigation strategies could be framed as shown in Fig. 7.15–increasing vehicle efficiency, using cleaner fuels and cleaner vehicles, reducing the activity data on car usage and mileages by improving public transport infrastructure and enhancing people's behavioural changes and finally putting every factor into a smart mobility model.

Based on IPCC report (IPCC 2014a, b), the technical potential exists to substantially reduce the current CO_2 emissions per passenger or tonne kilometre for all modes by 2030 and beyond. Energy efficiency and vehicle performance improvements range from 30% to 50% relative to 2010 depending on mode and vehicle type. Realizing this efficiency potential will depend on large investments by vehicle manufacturers, which may require strong incentives and regulatory policies in order

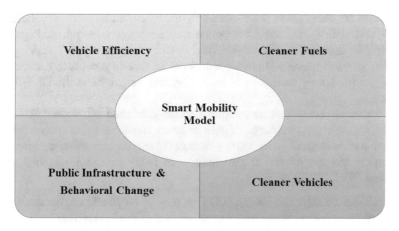

Fig. 7.15 Low-carbon transportation strategies

to achieve GHG emission reduction goals. Over the medium term (up to 2030) to long term (to 2050 and beyond), urban (re)development and investments in new infrastructure linked with integrated urban planning, transit-oriented development and more compact urban form that supports cycling and walking can all lead to modal shifts. Such mitigation measures could evolve to possibly reduce GHG intensity by 20–50% below 2010 baseline by 2050. In the following part of this section, different approaches will be elaborated one by one.

7.4.2   Vehicle Efficiency

Vehicle efficiency is vehicle's fuel efficiency, a measure of how far a vehicle could travel per unit of fuel. It is expressed in km/l or in miles per gallon (mpg) in the USA. More efficient vehicle means it could travel a longer distance given same amount of fuel or it could use less fuel for a given distance, resulting in less emission. Vehicle efficiency is a characteristic of the vehicle, which is dependent on the efficiency of the engine or propulsion system, the weight of the car and its aerodynamics (King 2007). It could also depend on the driver's driving behaviours, which will be further discussed in Sect. 7.4.4. Here we focus on vehicle itself.

According to King Review, technology for conventional vehicles that can reduce CO_2 emissions per car by 30% is already close to market, and this includes direct injection, variable valve actuation, cylinder deactivation, stop-start, regenerative braking and some non-propulsion technologies (King 2007). Although some of these technologies will carry a cost premium, higher fuel efficiency could reduce the total life cycle cost. King also mentioned barriers delaying deployment which include as follows: from the supply side, car manufacturers need to be confident of the market, and from the demand side, consumers want to pay a premium to have these vehicles.

Policies are being put in place to achieve dramatic improvements in vehicle efficiency, stimulating automotive companies to make major investments. Regulatory standards focused on fuel consumption and GHG emissions vary in their design and stringency. Some strongly stimulate reductions in vehicle size (as in Europe), and others provide strong incentives to reduce vehicle weight (as in the USA) (IPCC 2014a, b).

In the most recent fuel consumption guide published by Natural Resources Canada, it gave out very clear guidelines on how a consumer could choose the right car from a wide range of powertrains – such as the engine, transmission, drive shaft, suspension and the wheels. It also introduced today's technologies in Canada, including (NRC 2019):

- A cylinder deactivation system (CDS) in a 6- or 8-cylinder engine shuts down half of the cylinders when only a small amount of the engine's power is needed, which could save fuel consumption by 4–10%.
- Turbochargers force air into an engine's cylinders – unlike a standard engine, which draws air in at atmospheric pressure. This means that a smaller, turbocharged engine can produce the same power as a larger standard engine – and can lower fuel consumption by 2–6%.
- Variable valve timing (VVT) and lift systems adjust the timing of the engine valves to improve efficiency over a wide range of engine operating speeds. That leads to better operation of the engine and a 1–6% fuel reduction.
- Idle stop-start systems lower fuel consumption 4–10% or more and exhaust emissions by turning off the engine when the vehicle is idling and during deceleration at low speeds.
- Direct fuel injection increases engine's combustion efficiency because of a higher level of precision over the amount of fuel injected into the cylinder, the timing of the injection and the spray pattern. Direct injection can lower fuel consumption by 1–3%.

EnerGuide Label for vehicles in Canada uses a combined fuel consumption rating and separate city and highway fuel consumption ratings in L/100 km. The combined rating reflects 55% city and 45% highway driving. The 2019 Fuel Consumption Guide gave a comprehensive list with information about the fuel consumption of 2019 model year light-duty vehicles, where consumers could use this information to compare vehicles when shopping for the most fuel-efficient vehicle (NRC 2019).

7.4.3   Cleaner Fuels

Fuel CO_2 efficiency depends not only on the CO_2 released when the fuel is used or burnt in the engine (tailpipe emissions) but also the CO_2 emitted along the entire life cycle. As shown in Fig. 7.16, carbon emissions could occur at different stages of a fuel throughout its life cycle from the "well" (primary energy source) to the "wheel" (use in cars).

Fig. 7.16 Carbon emissions from fuel at different stages of its life cycle

For cars using conventional diesel or gasoline, normally 85% of emissions are from the cars' tailpipes, which is called downstream emission, while using alternative fuels, most emissions are from upstream, e.g. for biofuels, most emission from harvesting and production, and for electricity and hydrogen, emission almost totally from production (King 2007). The calculation of carbon emission from biofuels has been discussed in Chap. 2.

It should also be noted that emission is dependent on the different ways of making the fuels under study. For example, the following lists out the possible emission (King 2007):

- Petrol from oil sands has 25% higher carbon emission than from conventional sources.
- Biofuel carbon emissions vary between 10% and 100% of petrol emissions.
- Electricity varies between 5% and 90% depending on energy source.
- Hydrogen varies between 5% and 400% depending on energy source.

It basically says that although there is no emission from tailpipe of the new and low-carbon fuels, their upstream emission, the way to produce them, say, hydrogen, could be even four times compared with conventional fuels. Therefore, a life cycling evaluation of such fuels is essential in understanding the real emission reduction impact.

7.4.4 Cleaner Vehicles

Based on King Review, biofuel could only be part of the solution, and increases in vehicle efficiency of conventional vehicles will not be enough to get us where we need to be by 2050. In the long term, carbon-free road transport fuel is the only way to achieve decarbonized road transport, which includes electric vehicles with novel batteries charged by zero carbon electricity and hydrogen cars produced from zero carbon electricity.

The Facts and Figures
Electric and hybrid powertrains with clean electricity source are a promising technology in reducing the carbon emission and other road-side pollutants from the tailpipes. Kromer and Heywood calculated the carbon emission and energy use of the potential powertrains, such as gasoline hybrid-electric vehicles (HEVs),

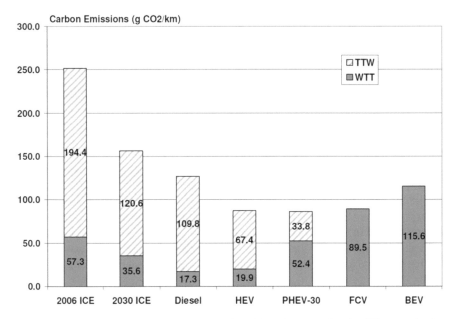

Fig. 7.17 Well-to-tank (WTT) and tank-to-wheel (TTW) carbon emission for different technologies (Kromer and Heywood 2007)

plug-in hybrid-electric vehicles (PHEVs), fuel-cell vehicles (FCVs) and battery-electric vehicles (BEVs) with a 2006 2.5L Toyota Camry as the basis for the future projections. The emissions of well-to-wheel (WTW) life cycle were divided into two stages: well-to-tank (WTT) and tank-to-wheel (TTW) (Kromer and Heywood 2007). A summary result is shown in Fig. 7.17. Compared with internal combustion engine (ICE) vehicles, the new powertrains could reduce the emission from its fuel's life cycle, especially in the TTW stage.

Figure 7.18 shows the WTW emission from different electricity sources. It also verified the above point that although no emission from TTW for FCVs and BEVs, their GHG reduction potentials are constrained by continued reliance on fossil fuels for producing electricity and hydrogen.

However, the above figures do not include the energy used to manufacture and recycle a vehicle, which is an important consideration when characterizing the total vehicle life cycle energy use and its corresponding carbon emissions. It was modelled that vehicle embodied carbon was around 20% of total emission (Kromer and Heywood 2007). For the BEV, the base vehicle (i.e. excluding the powertrain) accounts for 54% of life cycle emission of a vehicle, the lithium-ion battery for 26% and the remainder of electric powertrain for 20% (Nordelöf et al. 2014).

Besides carbon emission, study by Hawkins et al. showed that electric vehicles had the potential for significant increases in human toxicity, freshwater eco-toxicity, freshwater eutrophication and metal depletion impacts, largely emanating from the vehicle supply chain (Hawkins et al. 2012).

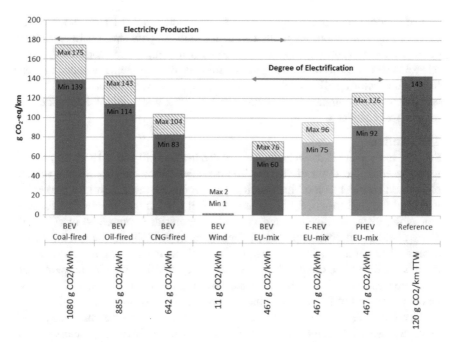

Fig. 7.18 WTW GHG emissions for different electricity sources and degrees of electrification (Nordelöf et al. 2014)

The Debates

According to McKinsey, electric vehicle sales are expected to exceed 100 million by 2035, and by 2050 there could be more than 2 billion electric vehicles on the road. The number of electric passenger car sales is expected to grow by more than a factor of 60 from 2018 to 2050 (McKinsey 2019). Could this really bring us to decarbonized road transport?

The Tesla Model S is an electric five-door liftback car, produced and introduced by Tesla Inc. on 22 June 2012. It was featured with quickest acceleration and longest range with a lower carbon emission.

However, in March 2016, one Tesla Model S owner in Singapore was reported that instead of getting rebate for low-carbon emission cars, he was required to pay $10,850 carbon emission surcharge on the vehicle (CNBC 2016). Tesla has clarified later on its website that for Tesla Model S, energy consumption rated at 181 Wh/km. In Singapore, given electricity generation released roughly 0.5 kgCO$_2$/kWh, carbon emission of Model S is 90 g/kM, while driving an equivalent gas-powered car like the Mercedes S-Class 500 results in emissions of approximately 200 gCO$_2$/km. But the test conducted by Singapore's transport authority found that that 2014 Model S's carbon emission was 222 gCO$_2$/km. Which data was flaw? This is still under debate, and the issue has not been resolved as till this moment Tesla S is not officially sold in Singapore and there are no superchargers around.

Just 1 month later after Singapore's case, an article titled "Electric shock – Tesla cars in Hong Kong more polluting than petrol models" published on 13 April 2016 on *South China Morning Post* stated that electric vehicles in Hong Kong could release 20% more carbon than regular petrol ones over the same distance in Hong Kong after factoring HK's coal-fired energy mix and battery manufactory. The data was based on the compared emissions over 150,000 km for a BMW 320i using petrol and Tesla Model 3.

The Solution

So, how to evaluate if the vehicle is a real cleaner one? Figure 7.19 gives a framework, based on the above discussion. To assess the carbon emission from a vehicle, one should consider to combine two life cycle processes: the life cycle of the vehicle itself and the well-to-wheel life cycle of the fuel used in the vehicle. The function unit is the distance travelled, and the carbon footprint is presented in gCO_2/km. There could be different approaches, depending on the "object" chosen for the life cycle analysis. If the fuel is chosen as the main object in study, the life cycle assessment is conducted on the fuel from well to wheel, which vehicle's emission could be treated as embodied carbon and incorporated at the "use" stage. Vice versa, embodied carbon of fuel could be incorporated in the "use and maintenance" stage of the vehicle life cycle. The "use" stages from these two processes are same to each other. By using this combined life cycle assessment, different parameters such as the electricity generation fuel mix, the production of the fuels, the production of batteries and disposal and recycling of the vehicle and its parts could be all reflected in the assessment.

7.4.5   Public Transport and Behavioural Change

Based on Eq. (7.1) described in Sect. 7.4.1, activity data is another parameter which affects the final transportation emissions. Activity data here is defined as total distance travelled (km/year) or total fuel used (l/year), which is influenced by behavioural issues like travel mode chosen or driving habits. Reduction actions could be considered from the following aspects:

- Behavioural change

  - Purchasing behaviour
  - Eco-driver

- Public infrastructure

  - Make bicycles a proper mode of transport
  - Public transportation options
  - Charging facilities

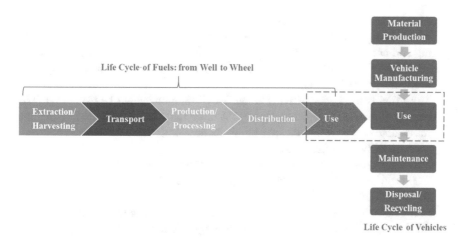

Fig. 7.19 Framework of combined life cycle assessment of fuels and vehicles

Behavioural Change

Adoption of the highly efficient or clean vehicles and realization of their carbon reduction measures depend on what vehicles the users choose and how they drive them.

According to King Review, choosing the most fuel-efficient model in the range can reduce a driver's carbon emissions by 25%, and driving a diesel engine of comparable performance, rather than size, can reduce a driver's emissions by around 15% (King 2007). Policies and incentives should be in place to encourage users to purchase low-carbon vehicles. For instance, in Hong Kong, the first registration tax (FRT) had been waived in full for electric private cars since 1994. But in order to discourage car ownership and to promote public transportation to reduce the road traffic congestion, the government changed its policy in 2017 that the FRT concession for EVs was capped at HK$97,500, which resulted in a nearly 97% drop in newly registered private EVs from 1,560 between April to October in 2016 to 49 during the same period in 2017. For Tesla alone, the number of newly registered cars for the full budget year dropped from 5,596 in 2016 to 50 in 2017. Government turned out to be wrong, as many users changed back to purchase petrol cars. Then in 2018, electric car buyers were given a HK$250,000 tax break if they traded in a car at least 6 years old. The tax concession would cover the FRT of private electric cars valued at HK$377,500 or less, which led to a rising demand for inexpensive EVs.

As how to drive, some smarter driving tips, summarized from King Review and Energy Saving Trust UK (energysavingtrust.org.uk), are listed below:

- Keep tyres pumped up.
  Under-inflated tyres increase the resistance when car is moving, resulting in a higher fuel consumption and more carbon emission. But over-inflated tyres can be unsafe. Drivers should regularly check the tyres and maintain it at a reasonable pressure.

- Accelerate and brake smoothly.
  Maintain a greater distance from the vehicle in front to avoid unnecessary braking and acceleration. Keep an eye on the traffic ahead and slow down early by gently lifting foot off the accelerator while keeping the car in gear.
- Moderate speed.
  More fuel is used when driving faster. At 70 mph one could be using up to 9% more fuel than at 60 mph and up to 15% more fuel than at 50 mph. Driving at a reasonable speed within the speed limit reduces emission.
- Shift up early to a higher gear.
  Driving at lower revs reduces fuel consumption so change up a gear at around 2,000 rpm for a diesel car and 2,500 rpm for a petrol car.
- Switch off engine.
  Many newer cars automatically turn off when stationary in neutral. If not, switch off the engine as fuel is wasted when the engine is idling.
- Minimize weight.
  Remove any excess items from the car to reduce the load. This will reduce the engine's workload and burn less fuel.
- Keep windows open.
  Air conditioning can increase fuel consumption by as much as 5%. It is more fuel efficient to open windows than using air conditioning when driving.
- Public transportation.
  Try to choose walking, cycling or public transportation whenever appropriate. This is very related to the public infrastructure and facilities, which will be discussed later.

Public Infrastructure
Public transport provision and infrastructure for walking, cycling, cleaner fuel distribution, EV charging, etc. also play an important role in enabling low-carbon transport. In this session, I will use Hong Kong as an example to elaborate three areas, i.e. walkability, public transportation and EV charging facilities.

For a dense and compact metropolitan with high reliance on public transport, walking is an essential daily activity for most citizens in Hong Kong. Improving walkability could alleviate road air pollution by motorized transportation to increase public health, as well as reduce carbon emission. Based on Hong Kong's unique urban features such as its "high density, mixed land use, constant traffic and pedestrian flow, hilly topography, use of space and connection with public transport", Civic Exchange, an independent Hong Kong-based public policy think tank, together with UDP International, developed a walk audit checklist to assess the walkability in Hong Kong, and it included ten categories (Ng et al. 2016): (a) accessibility and connectivity to nearby destinations; (b) physical and visual permeability for easy way finding; (c) public realm amenities; (d) attention to human scale and density; (e) variety and diversity; (f) legibility and orientation; (g) streets treated as public spaces that require appropriate management; (h) microclimate and environment; (i) safety and security; and (j) integration with public transport. Following this, the government launched "Walk in HK" initiative in 2017 and started a

Table 7.7 Comparison of emission factors of different transportation types in Hong Kong (Leung et al. 2010)

Transportation	EF (g CO_2e/man-km)	EF (g CO_2e/HK$)
MTR	7.8	11.5
Bus	27.9	49.3
Minibus	63.1 (diesel)/64.8 (LPG)	91.9 (diesel)/94.4 (LPG)
Tram	27.4	21.0
Ferry	2,228	1,478

30-month consultancy study on how to enhance walkability in Hong Kong aiming at formulating planning and design standards based on pedestrian-first principle, covering four themes, i.e. connectivity, safety, walking experience and smart information. Updates could be found on walk.hk.

Dr. Glenn Frommer, former Head of Corporate Sustainability for MTR Corporation, used to give remarks at various events, saying "I am the greenest person in Hong Kong, as I am a vegetarian and I take MTR!" Yes, with over 5 million trips made in an average weekday, mass transit railway (MTR) is the cleanest public transport method in Hong Kong, referred to Table 7.7.

Lastly, the EV charging facilities. Incentive for consumers to purchase EVs is not enough. Building required infrastructure is equally important to promoting low-carbon driving. In Hong Kong, there were approximately 11,100 EVs supported by 1,800 public charging points in 2018. Assistant Professor at Department of Geography and Resource Management of Chinese University of Hong Kong Sylvia He commented the charging situation in Hong Kong (He 2018):

- The ratio of EVs to charging lot is 7, which is far from sufficiently attractive for existing and potential EV users.
- In approximately 130 quick public EV charging stations (a quick charge takes 1–2 hours, compared with 6–7 hours for a standard charge) in the city, close to 40% are specifically for Tesla cars, rather than for all EVs.
- While public EV charging facilities are found in government premises, parks or carparks, they are not open 24 hours a day.
- Some management companies of private buildings or corporations are unwilling to install EV charging facilities in their parking spaces, citing high costs.
- EV users are also frustrated by incomplete information on the location and usage rate of EV charging stations.

Insufficient charging lot, uneven spread of charging points, lack of supercharging facilities, accessibility of the charging facility and lack of information sharing and policy to support charging at private properties are all the problems EV users are currently facing in Hong Kong. Based on her study with international comparisons, Dr. He concluded that Hong Kong lags behind in terms of official EV policy and infrastructure. This is back to my repeated point that no low-carbon policy could be developed or promoted alone, especially for low-carbon transportation. In this case, HK government needs an integral and coherent low-carbon transport strategy.

7.4.6 Smart Mobility

The smart mobility concept mobility-on-demand (MOD) was first introduced to me in 2008–2009 by Ryan Chin, who was then a researcher under late Professor William J. Mitchell at Massachusetts Institute of Technology (MIT) Media Lab. Prof. Mitchell created the Smart Cities research group in 2003 at MIT Media Lab and developed projects which included various electric vehicles as well as MOD systems that enabled one-way vehicle sharing (Mitchell et al. 2010). After Mitchell passed away in 2010, Smart Cities group has grown into Changing Places research group under Kent Larson at the MIT Media Lab.

Set up on 18 January 2011, Intelligent City Consortium (ICC) was initially made up of the MIT Media Lab (USA), the London School of Economics Enterprise (UK) and InnoZ (Germany), together with the Basque entities Denokinn (Spain) and AFYPAIDA (Spain). The consortium also counted the support of Carbon Care Asia for its operations in Hong Kong. ICC provided a space for collaboration among the most prestigious international institutions in terms of sustainable mobility that was to counsel big urban centres in the planet in order to solve their growing problems of pollution, traffic and power supply. Under ICC, the feasibility study of MOD has been conducted in a few metropolitans such as Barcelona, Berlin, London, San Francisco, Singapore and Sydney, and we have also submitted the proposal in Hong Kong but failed to secure the research funding. As far as I know, ICC has been dissolved, like many pioneers in innovation and disruption, but some MIT projects have flourished under some ICC members, which I will talk later.

In more recent years, the MOD programme became a new US Department of Transportation (USDOT) initiative led by the Intelligent Transportation Systems (ITS) Joint Program Office (JPO) and the Federal Transit Administration (FTA). The mission of the programme was to enable and leverage advancements in technology and operations to create an environment where all travellers could have safe mobility options, ensuring reliable, informed and efficient travel in a multi-modal network that prioritizes individual, on-demand mobility. In 2016, FTA awarded US\$8 million in MOD Sandbox grants to 11 projects to demonstrate integrated MOD concepts and solutions in real-world settings.

So, what is MOD? MOD is a new model of sustainable and personal urban mobility that emphasizes enhancing mobility options for all users through the integration of on-demand modal services, public transportation, payment mechanisms, traveller incentives and an array of real-time information services (Shaheen et al. 2017). Benefits of MOD include:

- Shared ownership provides users fractional ownership that allows them access to any vehicle in the fleet whenever they please and for as long as they need.
- MOD addresses directly the problem of last and first mile demand and encourages public transit.
- MoD systems provide extremely high utilization rates of all system resources, including parking spaces and vehicles.
- It reduces the pollution due to congestion and reduces carbon emissions.

USDOT has developed a very comprehensive framework on the MOD eco-system, which envisions an integrated and multimodal transportation operations management approach that can integrate the supply-side marketplace and users' needs on demand sides and how they interact with various stakeholders, e.g. government, local authorities, public transit agencies, transportation operators, apps and mobile services providers and the public (Shaheen et al. 2017). In the following part of this session, I will discuss the essential components of MOD systems, namely:

- Specially designed vehicles
- Vehicle stacks and racks and charging infrastructures
- Smart management systems that connect everything

Lightweight EVs (LEVs)
MoD systems consist of fleets of lightweight electric vehicles (LEV) placed at electric charging stations distributed throughout a city. Smart Cities developed three LEVs for MoD systems including the CityCar, RoboScooter and GreenWheel Smart Bicycle. Each of these vehicles utilize in-wheel motor technology called "Robot Wheels" that incorporate electric drive, steering, suspension and braking inside the hub space of a wheel. They are foldable to save the parking space, and they are lightweight that save energy or here electricity to power them

- CityCar
  The CityCar, developed by the Smart Cities group at the MIT Media Lab, are lightweight electric cars with in-wheel motors. The prototype developed in 2009–2010 was around 1,000 lbs, which was lighter than two-thirds of Smart Fortwo (i.e. 1,609 lbs), around one-third of Toyota Prius (i.e. 2,932 lbs) and one-fourth of Ford Explorer (i.e. 3,959 lbs). CityCar could fold and stack like shopping carts at the supermarket. Instead of side doors, they had doors in front of them, which made them extremely compact and efficient in the use of urban space. They were simple and modular in their design, robust, inexpensive and easy to maintain. They utilized contactless induction charging in their parking spaces – much as electric toothbrushes recharge in their holders – so they did not need very long ranges or to carry around large numbers of batteries. CityCar has developed further and commercialized as Hiriko by the Hiriko Driving Mobility Consortium in the Basque Country of northern Spain under the leadership of AFYPAIDA, who was the original member of ICC as mentioned above.
- RoboScooter
  Developed by Smart Cities in collaboration with Industrial Technology Research Institute and Sanyang Motor in Taiwan, RoboScooter are also lightweight, folding, in-wheel-motor electric vehicles. These two-wheelers are smaller, lighter and less expensive and consume less energy than four-wheel cars. They are particularly suitable for use where weather conditions are good, individual transportation is the priority and urban and economic conditions are less favourable to automobiles.

- GreenWheel
  Integrated in-wheel motor and battery hub system, smart bicycles could be electrified and provide additional power above and beyond the power provided by the person pedalling.

Vehicle Stacks/Racks
Stacks and racks are the vehicle pickup and dropoff points in MOD systems. They need to be distributed sufficiently at the services areas which should be close and convenient to the trip origins or destinations such as public transit or combined with the existing service points, e.g. supermarkets, banks, etc. Stacks and racks are also the charging points for LEVs in MOD systems. Hence, power connection and charging infrastructure specifications should be considered for these pickup and dropoff locations.

Smart Management System
The MOD system allows users to simply swipe a smart card and take a vehicle from one location and drive it to another charging station. So far, the applicable LEVs have been developed and can be deployed at the built vehicle stacks and racks at different pickup and dropoff points with charging facilities. To realize the MOD system, there are still two questions that need to be solved:

- How can the user locate the vehicle pickup/dropoff points?
- How to make sure there are sufficient vehicles and parking lots available at these points?

This leads to a smart management system, which includes the smart infrastructure with IoT, location sensors, mobile apps and networks communication. Users could use their smartphone to locate the nearest pickup/dropoff points and know whether there are vehicles available or how long they need to queue for one. Vehicles could be tracked by their location sensors or the sensors from the driver's smartphone.

Big data from historical data are needed to understand the fleet movement; with a given stock of vehicles and parking spaces, the system operator must try to keep supply and demand of vehicles in optimal balance across the system, which is a crucial element in MOD systems. Ten years ago, when we tried to conduct the feasibility study of MOD in Hong Kong, based on late Prof Mitchell's innovation, we found these two questions the technical bottleneck at that moment. But now with the development of ICT, wide application of smartphones and various sensors and IoT and big data, all these breakthrough technologies could realize MOD system for urban and personal smart transport.

Autonomous Future
An autonomous car is a self-driving car or driverless car. It is a vehicle that is capable of sensing its environment and moving with little or no human input. Most global auto manufacturers are actively developing autonomous technology, including BMW, Ford, General Motors, Honda, Tesla, Toyota, Volkswagen and Volvo (MIT Technology Review Insights 2019). Technical giant Google started its journey towards autonomous vehicles in 2009, and by 2017 Google's fleet WAYMO had

completed 3 million miles driving within four US states (Faisal et al. 2019). In February 2019, the alliance of Nissan Motor, Renault and Mitsubishi Motors planned to join Google's camp for developing autonomous taxis and other services using self-driving vehicles (Nikkei 2019). By 2020, Audi, BMW, Mercedes-Benz and Nissan are expecting to have their autonomous vehicles in the market (Faisal et al. 2019).

The potential benefits of autonomous cars include:

- Increased safety
- Less traffic and more efficient road use
- Increased driver productivity
- Energy savings
- Reduced emissions

Regarding GHG emission, it was reported in a previous review that autonomous driving could reduce energy use up to 80% from platooning, efficient traffic flow and parking, safety-induced light-weighting and automated ridesharing, but it was also found that total mileage could be doubled or tripled (Greenblatt and Shaheen 2015).

Since most of the current autonomous vehicle concepts have a person in the driver's seat who utilizes the communication tools to select destinations or routes, the term automated is more accurate and used in research papers. SAE International revised and released J 3016-2018: *Taxonomy and Definitions for Terms Related to Driving Automation Systems for On-Road Motor Vehicles*, and in January 2019 it unveiled a new visual chart as shown in Fig. 7.20. It is designed to clarify and simplify its J3016 "Levels of Driving Automation" standard for consumers. The J3016 standard defines six levels of driving automation, from SAE Level Zero (no automation) to SAE Level 5 (full vehicle autonomy). It serves as the industry's most-cited reference for automated vehicle (AV) capabilities.

1. Level 0 – No Driving Automation
   The performance of the driver of the entire dynamic driving task (DDT), as in conventional automobiles.
2. Level 1 – Driver Assistance
   A driving automation system characterized by the sustained and operational design domain (ODD)-specific execution of either the lateral or the longitudinal vehicle motion control subtask of the DDT. It is expected that the driver performs the remainder of the DDT.
3. Level 2 – Partial Driving Automation
   Similar to Level 1, but characterized by both the lateral and longitudinal vehicle motion control subtasks of the DDT with the expectation that the driver completes the object and event detection and response (OEDR) subtask and supervises the driving automation system.
4. Level 3 – Conditional Driving Automation
   The driver will be ready to respond to a request to intervene when issued by the automated driving system (ADS).

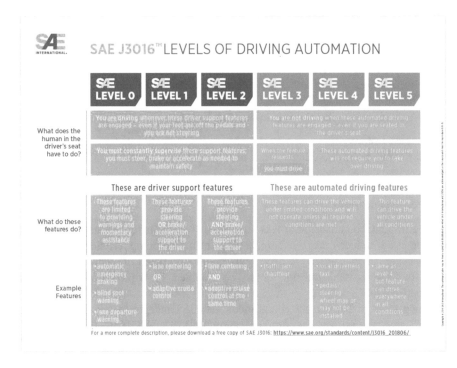

Fig. 7.20 SAE J3016 levels of driving automation

5. Level 4 – High Driving Automation
   Sustained and ODD-specific ADS performance of the entire DDT is carried out
   without any expectation that a driver will respond to a request to intervene.
6. Level 5 – Full Driving Automation
   Sustained and unconditional performance by an ADS of the entire DDT without
   any expectation that a driver will respond to a request to intervene.

It is expected that in near future, the convergence of automated vehicles and
MOD, so-called automated mobility on demand (AMOD), would have transformative impact on current infrastructures, personal mobility and goods delivery
(Shaheen 2018). Automated vehicles could solve major infrastructure problems,
and AMOD would further reduce the parking space, and the urban land use could
be better planned. With augmented safety features and limited self-driving ability,
the inclusion of level 3 automation into a carsharing fleet could decrease an operator's overall insurance costs, which could translate to cost savings for users. Level
4 or 5 automation would further enhance carsharing capability and attractiveness
(Greenblatt and Shaheen 2015). While AMOD become more mature in the future,
it could not only enhance the mobility access for all users but also could further
reduce the energy use and GHG emissions.

7.5   Resource Management

Post-consumer waste only contributes to less than 5% of GHG emissions, according to the IPCC report (Bogner et al. 2007). Globally, the largest source is landfill methane gas, followed by wastewater methane and nitrous oxide; in addition, minor emissions of carbon dioxide result from incineration of waste containing fossil carbon such as plastics and synthetic textiles. In Hong Kong, the total GHG emissions in 2016 amounted to 41.9 million tonnes CO_2e, among which electricity generation was the major source of emissions, amounting to 27.9 million tonnes or 66.5% of the total, and other major emission sources were the transport sector (17.9%) and waste management (5.9%).

However, in a life cycle view, material-related GHG is not only from the post-consumer waste management but covers emissions of all the life cycle stages from the raw material acquisition, production, transportation, consumption to end-of-life treatment. Based on OECD's study, although GHG emissions from the waste sector typically accounted for 3–4% of total emissions in OECD member countries' GHG emission inventories, when viewed from a life cycle perspective, GHG emissions arising from material management activities were estimated to account for more than half of total GHG emissions (ranging from 54% to 64%) for four analysed OECD member countries, which suggested a significant opportunity to potentially reduce emissions through modification and expansion of materials management policies (OECD 2012).

Carbon management on process and product level has been discussed in Chap. 6, and strategies such as business process reengineering and eco-design have been introduced as well. Hence in this section, we will start with the GHG related to the material life cycle and then focus on GHG emissions and mitigations of post-consumer waste management. Recycling, composting, combustion, landfilling and anaerobic digestion will be discussed in detail.

It should be noted that instead of waste management, the term resource management is used here, which is to emphasize two points. On the one hand, we should conserve our resources by use less and waste less. On the other hand, waste could be used as a valuable resource to replace the virgin materials for production and to generate energy to avoid fossil fuel use. Waste management should be viewed in holistic resource management.

7.5.1   GHG Related to Material Life Cycle

Material life cycle stages, as shown in Fig. 7.21, include raw material extraction and processing, products manufacture, product use by consumers, waste management of end-of-life products and transportation of materials and products in between. GHG

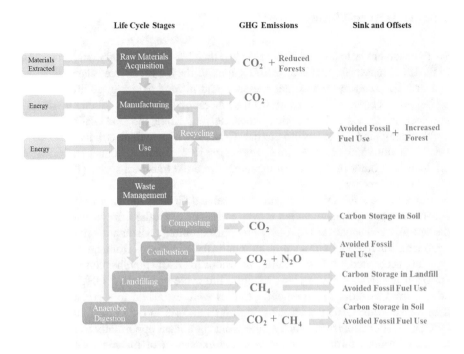

Fig. 7.21 Carbon emissions in materials life cycle

related to material life cycle can be analysed and categorized into GHG emissions and carbon sinks. GHG emission sources along the material life cycle include:

- Reduced carbon sink such as forests when using virgin materials like paper
- Energy-related carbon emission, such as electricity or fuels consumed in raw material processing, product manufacture, compost machinery and transportation
- Non-energy-related carbon emission, such as CO_2 released in steel manufacturing when limestone is converted to lime
- Nonbiogenic CO_2, N_2O emissions from waste combustion
- Uncontrolled CH_4 emission from landfills
- Uncontrolled biogas ($CO_2 + CH_4$) emission from anaerobic digestion

  Meanwhile, carbon sink and offset along the material life cycle include:

- Forest carbon sequestration
- Soil carbon storage in compost and landfills
- Avoided fossil fuel emission such as utilization of landfill gas and biogas to generate electricity to avoid utility emission on-site

7.5.2 Reduction, Recycling and Upcycling

Source Reduction

Source reduction, also known as waste prevention, is the elimination of waste before it is created. It is measured by the amount of product that would otherwise be produced but is not generated. GHG emissions associated with making the product and managing the postconsumer waste are avoided. Source reduction has:

1. Avoided GHG emission from raw material acquisition
2. Avoided GHG emission from product manufacturing
3. Increased forest carbon sequestration in the case of paper products
4. Zero waste management GHG emissions

Recycling

Instead of being disposed of, a product can be recycled and used in place of virgin materials in the manufacturing process. The avoided GHG emissions from remanufacture using recycled inputs is calculated as the difference between the GHG emissions from manufacturing a material from 100% recycled inputs and the GHG emissions from manufacturing an equivalent amount of the material from 100% virgin inputs, including the process of collecting and transporting the recyclables (USEPA 2006). No GHG emissions occur at the waste management stage because the recycled material is diverted from waste management facilities.

Example #1 Reduction Versus Recycling

Given:

GHG emissions of waste management options compared to landfilling (tCO_2e/t) (management option net emission minus landfilling net emission) in Table 7.8.

Question:

What are the avoided carbon emissions if we recycle 100 tonnes of office paper? How about if we reduce our consumption of 100 tonnes of office paper?

Answer:

Reduced emissions by recycling = $100 \times 1.31 = 131$ tCO_2e
Use less office paper = $100 \times 2.71 = 271$ tCO_2e

With same amount of paper, compared with recycling, more carbon emissions would be avoided by source reduction.

Table 7.8 GHG emissions of office paper (USEPA 2006)

Material	Source reduction (100% virgin inputs)	Recycling
Office paper	−2.79	−1.31

Upcycling

Upcycling is often considered as a process in which waste materials are converted into something of higher value and/or quality in their second life. Although there is no common definition of upcycling, in a recent review (Sung 2015), it was found that the most definitions of upcycling in research paper originated from *Cradle to Cradle: Remaking the Way We Make Things*, a 2002 book by German chemist Michael Braungart and US architect William. It called for a radical change in industry to switch from a cradle-to-grave pattern to a cradle-to-cradle pattern and advocated innovations for perpetually circular material reutilisation as opposed to current recycling practice. Based on their vision of upcycling, when products reach their end of life, they become either "biological nutrients" or "technical nutrients". Biological nutrients are materials that can re-enter the environment. Technical nutrients are materials "that is designed to go back into the technical cycle, into the industrial metabolism from which it came. [...] Isolating them from biological nutrients allows them to be upcycled rather than recycled – to retain their high quality in a closed-loop industrial cycle" (Braungart and McDonough 2002).

Built on traditional idea of recycling, upcycling is an innovative process that aims to achieve better quality and performance through better product design. Upcycling has been increasingly recognized as a promising means to reduce material and energy use, and compared with traditional recycling, it could have less carbon emission; low-carbon concept and reduced downstream footprint are embedded into the design concept.

7.5.3   Composting

Composting is an aerobic process that decomposes degradable organic carbon in the waste material into CO_2, water and a humic fraction, and some carbon could be stored in the residual compost. Biogenic CO_2 emissions associated with decomposition, both during the composting process and after the compost is added to the soil, are not counted in the GHG inventory based on IPCC guidelines (IPCC 2006). Potential GHG emissions from composting process include:

- CH_4 formed in anaerobic sections of the compost. It was found by USEPA that the well-managed compost process in a good aerobic environment with proper moisture control normally does not generate CH_4. Even CH_4 is formed in anaerobic sections of the compost, it is most likely oxidized to CO_2 in the aerobic sections of the compost (USEPA 2006). The estimated CH_4 released into the atmosphere ranges from less than 1% to a few percent of the initial carbon content in the material (Bogner et al. 2007).

- Energy-related carbon emissions including fuels used for transportation of feed-stock materials and compost, electricity and other energy used for machinery such as vessels, drums or other machines used for turning the compost piles or for aeration.

Depending on compost quality, compost could be used as organic fertilizer in agriculture or soil enhancer for soil stabilization and soil improvement with its increased organic matter and higher water-holding capacity. As studied by USEPA, potential carbon storage of compost includes (USEPA 2006):

- Some carbon content in the compost is retained by the soil system, when adding compost into the soil to increase organic matter inputs.
- Nitrogen in compost can stimulate higher productivity, thus generating more crop residues, which increase soil carbon.
- The composting process leads to increased formation of stable carbon compounds (e.g. humic substances, aggregates) that then can be stored in the soil for long (>50 years) periods of time.

The net carbon emission from composting process is the subtraction difference of the above potential GHG emissions and carbon storage. USEPA (2006) and IPCC (2006) published different equations and models to estimate the net carbon emissions from the compost, depending on degradable organic carbon content and C/N ratio of treated feed stock, compost site operation, compost application, etc. Interested readers could refer to the above two references for further reading.

7.5.4   Combustion

Waste incineration is defined as the combustion of solid and liquid waste in controlled incineration facilities. Types of waste incinerated include municipal solid waste (MSW), industrial waste, hazardous waste, clinical waste and sewage sludge. These processes include incineration with and without energy recovery, production of refuse-derived fuel (RDF) and industrial co-combustion. Incineration has been widely applied in many developed countries, especially those with limited space for landfilling such as Japan and many European countries.

Combustion of waste will release GHGs including CO_2, CH_4 and N_2O. Normally, emissions of CO_2 from waste incineration are more significant than CH_4 and N_2O emissions, and combustion of MSW results in emissions of CO_2 (because nearly all of the carbon in MSW is converted to CO_2 under optimal conditions) and N_2O (USEPA 2006). It should be noted that the climate-relevant CO_2 emissions from waste incineration are determined by fossil carbon content of the waste. The proportion of carbon of biogenic origin is usually in the range of 33–50%. The allocation to fossil or biogenic carbon has a crucial influence on the calculated amounts of climate-relevant CO_2 emissions (Johnke 2000). It should also be noted that traditional air pollutants from combustion such as non-methane volatile organic compounds (NMVOCs), carbon monoxide (CO), nitrogen oxides (NOx) and sulphur oxides (SOx) are of major concern to public health, but not counted in the GHG inventory.

Incineration reduces the mass of waste and can offset fossil fuel use. GHG emissions are avoided at facilities with energy recovery in a waste-to-energy (WTE) plant. The electricity produced by a WTE plant replaces electricity that would otherwise be provided by a local utility power plant, where fossil fuels such as coal and natural gas are normally burned to generate electricity. The utility carbon emission is then avoided. The avoided utility emission is affected by three factors: (1) the energy content of mixed waste and of each separate waste material stream, (2) the combustion system efficiency in converting energy in waste to delivered electricity and (3) the emission conversion factor ($kgCO_2e/kWh$) of the local electric utility plant. The net GHG emission from waste combustion is the avoided emission subtracted by the emissions from the combustion.

T · PARK – Sewage Sludge Incineration Plant in Hong Kong
Burning is a highly effective method of sewage sludge treatment. Advanced incineration technology through the high-tech thermal process is adopted to ensure efficient and reliable treatment of sludge. With two plants of four incineration trains in the facility, T · PARK can handle a maximum capacity of 2,000 tonnes of sludge per day.

The heat energy generated from the incineration process is recovered and turned into electricity that can support the needs of the entire facility. When running at full capacity, it can produce up to 2 megawatts (MW) of surplus electricity for the public power grid (can support 4,000 household), which is an impeccable example of "waste-to-energy" in action.

After incineration, sludge will be converted into ash and residues – a total reduction of 90% of the original sludge volume. This dramatically cuts down the quantity of waste to be disposed of in the landfills and reduces the emission of greenhouse gases by up to 237,000 tonnes a year.

Source: www.tpark.hk

7.5.5   Landfilling

Sanitary landfilling has been the most commonly used waste disposal method, especially for MSW, and according to IPCC, landfill methane is the largest contributor to GHG emissions of post-consumer waste management (Bogner et al. 2007). Figure 7.22 illustrates the organic decomposition processes and carbon mass balance in landfill.

When waste is dumped into the landfill site, the organic matter in the waste is first decomposed by aerobic microorganisms until oxygen is consumed. CO_2 is produced in this initial stage. Anaerobic decomposition is then started with the

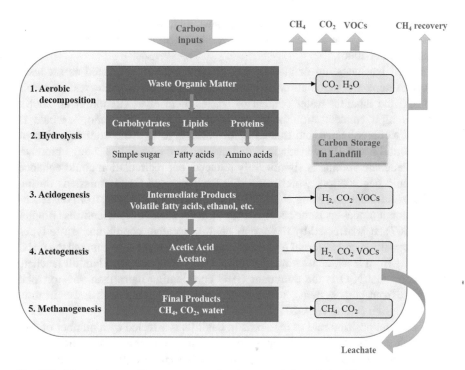

Fig. 7.22 Waste decomposition processes and carbon mass balance in landfills

hydrolysis process, through which the complex organic molecules are broken down into simple sugars, amino acids and fatty acids. The next stage in the anaerobic process is acidogenesis, where organic compounds are further broken down and volatile fatty acids, ammonia and CO_2 are produced. Simple molecules are further digested by acetogenic bacteria to produce acetic acid, CO_2 and H_2. The final stage of anaerobic decomposition is methanogenesis, where all the intermediate products of the above processes are converted to CH_4, CO_2 and water, which make up the majority of landfill gases. Once the methanogenic stage begins, landfill gas generated is composed of approximately 50% CH_4 and 50% CO_2. But landfill gas as collected generally has a higher CH_4 concentration than CO_2 concentration (sometimes as much as a 60:40 ratio), because some of the CO_2 is dissolved in the leachate as part of the carbonate system ($CO_2 \leftrightarrow H_2CO_3 \leftrightarrow HCO_3^- \leftrightarrow CO_3^{2-}$) (USEPA 2006).

As shown in Fig. 7.22, in view of the landfill as a system, inputted carbon could be released from the system as the gases, such as CO_2, CH_4 and VOCs; it can also be released with the outflow of the leachate from the system, while the remaining carbon could be stored in the system.

VOCs released from landfills may include saturated and unsaturated hydrocarbons, acidic hydrocarbons, organic alcohols, aromatic hydrocarbons, halogenated compounds, sulphur compounds and mercaptans, and the total amounts of those VOCs are usually below 1% (by volume) of the total landfill gas emissions (Saral et al. 2009). USEPA also reported that VOCs would have a small role in the overall carbon balance, as concentrations of CH_4 and CO_2 will both be hundreds of times larger (USEPA 2006).

Landfill leachates are defined as the aqueous effluent generated as a consequence of rainwater percolation through wastes, biochemical processes in waste's cells and the inherent water content of wastes themselves (Renou et al. 2008). With its high concentration in ammonia and organic compound, leachate is required to be collected and treated before discharge to avoid contamination of underground water. If the leachate is recirculated to the landfill as a large bioreactor, the leachate volume is significantly reduced, but recirculation could enhance the organic degradation, which results in the increased GHG emissions during leachate recirculation (Wang et al. 2014). CH_4 and N_2O produced from the leachate treatment processes is the second largest GHG emission from landfills (Bogner et al. 2007). It is affected by the age of landfill, weather conditions, waste types and composition and different treatment processes. Wang et al. compared the GHG emissions from leachate treatment systems with young and aged landfill leachate that found that N_2O is the dominant GHG contributing more than 95% of total emission from both systems, while system with aged leachate is lower in total GHG emission (Wang et al. 2014).

The decomposition rate of materials in landfills is affected by a number of factors, such as waste type and composition, moisture, temperature and microorganisms. Organic matter contains varying amounts of cellulose (the main component of all plant tissues and fibres), hemicellulose (the constituent of plant material that binds with cellulose to form the network of fibres) and lignin (the integral part of the cell wall that fills space between cellulose and hemicellulose). While cellulose and hemicellulose easily biodegrade, lignin does not. In addition, the presence of lignin can also prevent the decomposition of cellulose and hemicellulose. Materials with high lignin content (such as newspapers and branches) have higher landfill carbon storage potential than materials with lower lignin content (such as food scraps) (USEPA 2010).

The GHG inventory from landfill is then considered as follows:

- CO_2 from organic decomposition in landfill: It is not counted as if not in landfills it would also be decomposed in a natural environment.
- CH_4 from landfill gas: It is the major GHG source from landfills and CH_4 is counted as an anthropogenic GHG, because even though it is derived from sustainably harvested biogenic sources, degradation would not result in CH_4 emissions if not for deposition in landfill.
- Carbon reduction or avoided utility emission from CH_4 recovery: CH_4 could be recovered by flaring or generation of heat and electricity from it.
- CH_4 and N_2O emissions from leachate treatment facility: If leachate is recirculated, the increased emission is already counted in CH_4 emission; if it is treated separated in a treatment facility on-site or off-site, both emissions should be calculated.
- Energy-related emission: emissions from fuel or electricity used for transportation of waste and materials and machinery used on site should be covered.

- Carbon storage: Organic matter could not be decomposed completely, and the remaining carbon is stored in the landfill. Because this carbon storage would not normally occur under natural conditions, this is counted as a carbon sink. However, fossil carbon such as plastics that remains in the landfill is not counted as stored carbon (USEPA 2006).

7.5.6   Anaerobic Digestion

Anaerobic digestion is a process which breaks down organic matter into simpler chemical components in lack of oxygen. The decomposition process is illustrated in Fig. 7.22 after aerobic digestion in landfills. Anaerobic digestion is applicable to treat organic waste such as sewage sludge, food waste, manures, green and yard waste, etc.

As shown in Fig. 7.23 (the diagram in the blue square), the collected organic waste is first sent to the pretreatment facility, where the feedstock is better mixed, the moisture content is adjusted and undesirable materials such as large items and inert materials, e.g. plastic and glass, are removed as the rejected materials, which are normally disposed at landfills. The rejection rate depends on the type and composition of the waste treated. Pretreated organic waste is then sent to the digester, where anaerobic digestion takes place. Digesters can be classified as dry or wet, as well as in relation to the temperature and the number of stages, to suit for different types of organic waste with optimum operating conditions.

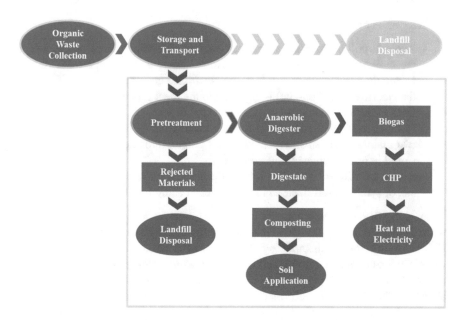

Fig. 7.23   A typical anaerobic digestion process

After digestion, the organic matter is broken down by the microorganisms, and biogas is released, and digested solid is left behind as digestate. Biogas can be purified by removing the CO_2 and water vapour and then used in a CHP unit to generate heat and electricity. The by-product digestate can be directly applied to agriculture lands or be further processed into compost to increase its quality as an organic fertilizer for agriculture applications.

The GHG emissions and avoidance related to anaerobic digestion cover the following:

- GHG emissions from

  - Transport of waste materials.
  - Pretreatment and digester operations. Pretreatment process normally includes grinding, screening and mixing the feedstock before they are fed into the reactor, where electricity is consumed. In the wet digester, digestate is dewatered and electricity is used in the dewatering process. For the dry digester, the digestate is removed without dewatering; however, more diesel is needed for dry digestion operations as it involves the additional use of front-end loaders to move materials.
  - Treatment and transport of rejected materials.
  - Biogas collection and utilization. Biogas cannot be fully utilized, and around 2% CH_4 is leaked during the digestion process, according to USEPA's Waste Reduction Model (WARM) (USEPA 2019).
  - Digestate composting. Fugitive emissions of CH_4 and N_2O are detected during the curing process (USEPA 2019).
  - Transport of compost and soil application.

- GHG sink or avoidance from

  - Avoided utility emission. Generated heat and electricity from biogas offset the emission from fossil fuel combustion on-site and the utility emissions. Normally surplus electricity (i.e. the electricity generated from CHP minus the electricity used on-site) is exported to the local grid.
  - Carbon storage. Similar to what is discussed in Sect. 7.5.3 of compost, carbon from digestate applied to agricultural lands remains stored in the soil through two main mechanisms: direct storage of carbon in depleted soils and carbon stored in non-reactive humus compounds.
  - Avoided fertilizer. When the compost is applied on land, it could replace the use of synthetic fertilizer, which is produced using fossil fuels.

In 2018/2019, collaborating with UK's environmental consulting firm Eunomia, we at University of Edinburgh's Hong Kong Centre for Carbon innovation helped HKEPD to develop the methodology to calculate the carbon reduction of the Hong Kong's first anaerobic digestion facility for food waste. After reviewing the different methodologies in the literatures and overseas cases such as US WARM mentioned above, UK's Emissions Performance Standard (EPS) developed by the Greater London Authority and the reporting scheme for food waste recycling used by the South Korean, the following emissions sources are therefore taken into account:

- Fuel use in transport systems used for waste collection
- Treatment process emissions, including those associated with rejected material or contaminants and those associated with energy used in the process
- Emissions associated with the management of digestate or compost (including those associated with dewatering, the use of the product on land and transport emissions)
- Avoided fossil emission by energy generation
- Avoided emission from compost application

It is noted that since the carbon reduction baseline is that food waste is sent to be disposed at landfill site in Hong Kong, the transportation of waste and rejected material are both counted, while in most case studies, they are out of the system boundary. Carbon storage and emission of compost transport and land application are not counted in this case.

O · PARK1 – First Organic Resources Recovery Centre in Hong Kong
O · PARK1, locating at Siu Ho Wan of North Lantau, adopts anaerobic digestion technology to convert food waste into biogas for electricity generation, while the residues from the process can be produced as compost for landscaping and agriculture use. O · PARK1 is capable of handling 200 tonnes of food waste per day. The biogas will be used to generate electricity, and apart from the internal use within O · PARK1, about 14 million kWh of surplus electricity, which is equivalent to the power consumption by some 3,000 households, can be exported each year. The detailed processes could be referred to its website: https://www.opark.gov.hk/en/process.php.

References

Arup (2018) Energy systems: a view from 2035. What will a future energy market look like? Energy System in 2035

BCA (2018) Super low energy building technology roadmap. Building and Construction Authority, Singapore

Bendell J (2018) Deep adaptation: a map for navigating climate tragedy. IFLAS Occasional Paper 2, 27 July 2018. www.iflas.info. Updated in December 2018

Bogner J, Abdelrafie Ahmed M, Diaz C, Faaij A, Gao Q, Hashimoto S, Mareckova K, Pipatti R, Zhang T (2007) Waste management. In: Metz B, Davidson OR, Bosch PR, Dave R, Meyer LA (eds) Climate change 2007: mitigation. Contribution of Working Group III to the Fourth Assessment Report of the Intergovernmental Panel on Climate Change. Cambridge University Press, Cambridge, UK/New York

Braungart M, McDonough W (2002) Cradle to cradle. Remaking the way we make things. Vintage, London, pp 109–110

ClimateWorks Australia, ANU, CSIRO, CoPS (2014) Pathways to deep decarbonization in 2050: how Australia can prosper in a low carbon world: technical report. Published by ClimateWorks Australia, Melbourne, Victoria

CNBC (2016) Tesla Model S owner protests Singapore's carbon emissions surcharge. Published 8 Mar 2016. Cnbc.com. Accessed 30 June 2019

Darko A, Zhang C, Chan APC (2017) Drivers for green building: a review of empirical studies. Habitat Int 60:34–49

Deep Decarbonization Pathways Project (2015) Pathways to deep decarbonization 2015 report – executive summary, SDSN – IDDRI

Denis A, Jotzo F, Ferraro S, Jones A, Kautto N, Kelly R, Skarbek A, Thwaites J, Adams P, Graham P, Hatfield-Dodds S (2014) Pathways to deep decarbonization in 2050 – how Australia can prosper in a low carbon world. The Australian report of the Deep Decarbonization Pathways Project of the Sustainable Development Solutions Network and the Institute for Sustainable Development and International Relations

EMSD (2015) Code of practice for building energy audit. Electrical and Mechanical Services Department

EMSD (2018) Hong Kong energy end-use data 2018. The Energy Efficiency Office, Electrical and Mechanical Services Department, Issued Sept 2018

Faisal A, Yigitcanlar T, Kamruzzaman M, Currie G (2019) Understanding autonomous vehicles: a systematic literature review on capability, impact, planning and policy. J Transp Land Use 12(1):45–72

Four Twenty Seven (2018) Climate risk, real estate, and the bottom line Four Twenty Seven and GeoPhy

Greenblatt JB, Shaheen S (2015) Automated vehicles, on-demand mobility, and environmental impacts. Curr Sustain Renew Energy Rep 2:74–81. https://doi.org/10.1007/s40518-015-0038-5

Haas H (2011) Harald Haas: wireless data from every light bulb. Harald Haas at TEDGlobal 201. Ted.com. Accessed June 2019

Hawken P (2017) Drawdown the most comprehensive plan ever proposed to reverse global warming. Penguin Books, New York

Hawkins TR, Singh B, Majeau-Bettez G, Strømman AH (2012) Comparative environmental life cycle assessment of conventional and electric vehicles. J Ind Ecol 17(1):158–160

He S (2018) Tax breaks alone won't promote e-mobility. Published on 14 June 2018. www.ejinsight.com. Accessed July 2019

Higuera J, Llenas A, Carreras J (2018) Trends in smart lighting for the Internet of Things Published in ArXiv. arXiv:1809.00986 [cs.CY]. 2018

Hui SCM (2009) Study of thermal and energy performance of green roof systems. Final report, Department of Mechanical Engineering, The University of Hong Kong, Hong Kong

IEA (2018) CO_2 emissions from fuel combustion 2018 highlights

IPCC (2006) Chapter 4: Biological treatment of solid waste IPCC/IGES. In: IPCC guidelines for national greenhouse gas inventories. Vol 5: Waste. Institute for Global Environmental Strategies (IGES), Hayama. http://www.ipcc-nggip.iges.or.jp/public/2006gl/ppd.htm

IPCC (2014a) Climate change 2014: synthesis report. Contribution of Working Groups I, II and III to the Fifth Assessment Report of the Intergovernmental Panel on Climate Change. IPCC, Geneva, 151 pp

IPCC (2014b) Transport. In: Climate change 2014: mitigation of climate change. Contribution of Working Group III to the Fifth Assessment Report of the Intergovernmental Panel on Climate Change. Cambridge University Press, Cambridge, UK/New York

Johnke B (2000) Emission from waste incineration. In: Report on good practice guidance and uncertainty management in national greenhouse gas inventories. Chapter 5, Section 3. The report was accepted by IPCC Plenary at its 16th session held in Montreal, 1–8 May 2000

King (2007) King review on low carbon cars. Part I: The potential for CO_2 reduction. October 2007. Crown Copyright

Kromer MA, Heywood JB (2007) Electric powertrains: opportunities and challenges in the U.S. light-duty vehicle fleet. Publication No. LFEE 2007-03 RP

Leung KH, Leung YC, Chan HS, Cheng D, Shek T, Lam CK, Cheng WK, Koo CP, Yun D (2010) Carbon audit toolkit for small and medium enterprises in Hong Kong. The University of Hong Kong. ISBN: 978-962-85138-7-1

Liu G, Shek LC, Lee SF, Li D (2013) Journey of energy efficient lighting technology and development of energy efficiency requirements of lighting installation in building energy code. In: 5th international lighting symposium 2013 on high quality and energy efficient lighting – realities & issues, 18 Oct 2013, Hong Kong

Liu Q, Chen Y, Tian C, Zheng X, Li J (2016) Strategic deliberation on development of low-carbon energy system in China. Adv Clim Chang Res 7:26–34

McKinsey (2019) Global energy perspective 2019: reference case. Summary. Energy Insight by McKinsey

MIT Technology Review Insights (2019) Self-driving cars take the wheel. Advanced technologies come together to get autonomous vehicles driving safely and efficiently. MIT Technology Review Insights, 15 Feb 2019. https://www.technologyreview.com/s/612754/self-driving-cars-take-the-wheel/. Accessed July 2019

Mitchell WJ, Borroni-Bird CE, Burns LD (2010) Reinventing the automobile: personal urban mobility for the 21st century. The MIT Press, Cambridge, MA

Morau D, Libelle T, Garde F (2012, 2012) Performance evaluation of green roof for thermal protection of buildings in Reunion Island. Energy Procedia, Elsevier (14):1008–1016. https://doi.org/10.1016/j.egypro.2011.12.1047

Murtagh N, Roberts A, Hind R (2016) The relationship between motivations of architectural designers and environmentally sustainable construction design. Constr Manag Econ 34(1):61–75. https://doi.org/10.1080/01446193.2016.1178392

Ng S, Lai C, Liao P, Lao M, Lau W, Govada S, Spruijt W (2016) Measuring and improving walkability in Hong Kong. Final report, Civic Exchange and UDP International

Nikkei (2019) Nissan-Renault alliance to join Google on self-driving cars. Automaker group goes all-in on outside partnerships for future of driving. By Furukawa K and Yamamoto N, Nikkei staff writers, 5 Feb 2019. https://asia.nikkei.com/Business/Companies/Nissan-Renault-alliance-to-join-Google-on-self-driving-cars. Accessed July 2019

Nordelöf A, Messagie M, Tillman AM, Söderman ML, Van Mierlo J (2014) Environmental impacts of hybrid, plug-in hybrid, and battery electric vehicles – what can we learn from life cycle assessment? Int J Life Cycle Assess 19:1866–1890. https://doi.org/10.1007/s11367-014-0788-0

NRC (2019) 2019 Fuel consumption guide. Natural Resources Canada. Updated on vehicles. nrcan.gc.ca

OECD (2012) Greenhouse gas emissions and the potential for mitigation from materials management within OECD countries. Working Group on Waste Prevention and Recycling. Environment Directorate, Environment Policy Committee, Organisation for Economic Co-operation and Development, 28 Mar 2012. ENV/EPOC/WGWPR(2010)1/FINAL

PRI (2016) Responsible investment market update: a snapshot of signatory action

PRI (2018) Impact investing market map

Renou S, Givaudan JG, Poulain S, Dirassouyan F, Moulin P (2008) Landfill leachate treatment: review and opportunity. J Hazard Mater 150:468–493

Saral A, Demira S, Yıldız S (2009) Assessment of odorous VOCs released from a main MSW landfill site in Istanbul-Turkey via a modelling approach. J Hazard Mater 168:338–345

Shaheen S, Cohen A, Yelchuru B, Sarkhili S (2017) Mobility on demand operational concept report. Produced by Booz Allen Hamilton, U.S. Department of Transportation Office of the Assistant Secretary for Research and Technology, Intelligent Transportation Systems Joint Program Office. Final report – September 2017. FHWA-JPO-18-611

Shaheen S (2018) Shared mobility: the potential of ride hailing and pooling. UC Berkeley: Transportation Sustainability Research Center. Retrieved from https://escholarship.org/uc/item/46p6n2sk 1 March 2018

Sung K (2015) A review on upcycling: current body of literature, knowledge gaps and a way forward. In: The ICECESS 2015: 17th international conference on environmental, cultural, economic and social sustainability, Venice, Italy, 13–14 Apr 2015

Teng F et al (2015) Pathways to deep decarbonization in China, SDSN – IDDRI

UNEP (2017) The Global Status Report 2017: towards a zero-emission, efficient, and resilient buildings and construction sector. Prepared by Thibaut Abergel, Brian Dean and John Dulac of the International Energy Agency (IEA) for the Global Alliance for Buildings and Construction (GABC). ISBN: 978-92-807-3686-1

UNEP (2018) 2018 Global Status Report: towards a zero-emission, efficient and resilient buildings and construction sector. Prepared by the International Energy Agency (IEA) for the Global Alliance for Buildings and Construction (GlobalABC). ISBN: 978-92-807-3729-5

USEPA (2006) Solid waste management and greenhouse gases. A life-cycle assessment of emissions and sinks, 3rd edn. U.S. Environmental Protection Agency, Washington, DC

USEPA (2010) Landfill carbon storage in EPA's waste reduction model, 27 Oct 2010. https://www.epa.gov/warm/landfilling-and-landfill-carbon-storage-waste-reduction-model-warm Accessed July 2019

USEPA (2019) Documentation for greenhouse gas emission and energy factors used in the Waste Reduction Model (WARM). Management practices chapters. Prepared by ICF For the U.S. Environmental Protection Agency Office of Resource Conservation and Recovery

Wang Y, Teter J, Sperling D (2011) China's soaring vehicle population: even greater than forecasted? Energy Policy 39:3296–3306

Wang X, Jia M, Chen X, Xu Y, Lin X, Kao CM, Chen S (2014) Greenhouse gas emissions from landfill leachate treatment plants: a comparison of young and aged landfill. Waste Manag 34:1156–1164

Williams JH, Haley B, Kahrl F, Moore J, Jones AD, Torn MS, McJeon H (2014) Pathways to deep decarbonization in the United States. The U.S. report of the Deep Decarbonization Pathways Project of the Sustainable Development Solutions Network and the Institute for Sustainable Development and International Relations. Revision with technical supplement, 16 Nov 2015

Index

© Springer Nature Switzerland AG 2020
S. W. W. Zhou, *Carbon Management for a Sustainable Environment*,
https://doi.org/10.1007/978-3-030-35062-8

Printed in the United States
By Bookmasters